Eco-Anxiety and Ecological Citizenship

"Michel Bourban has written a book that—finally—synthesizes and explains in exquisite conceptual and operational detail what we mean when we speak of "eco-anxiety." With the pressing planetary challenge of climate change driving our global anxiety about the future, this book arrives at just the right moment to both advance the dialectic and give us reason for hope. The writing is trenchant, yet accessible. It should be of value to scholars and students alike."

—John Allegrante, *Charles Irwin Lambert Professor of Health Behavior and Education, Teachers College, Columbia University, and principal co-editor, Anxiety Culture: The New Global State of Human Affairs, Baltimore, Johns Hopkins University Press, 2024*

"Michel Bourban provides a short and pithy introduction to an underexplored challenge of our time: how to understand eco-anxiety and engage with it productively together to make a better world. Bourban navigates these troubled waters with skill, sensitivity, insight, and heart. His preferred strategy—embracing ecological citizenship and a key set of "green" virtues—will resonate with and inspire many environmentalists. I see much to admire, some disagreements, and plenty of scope for fruitful engagement. In short, the book provides a stimulating entry-point into what have become vital conversations in our challenging times. Highly recommended for students, academics, and citizens of all kinds."

—Stephen Gardiner, *Ben Rabinowitz Endowed Professor of Human Dimensions of the Environment at the University of Washington, and editor of The Oxford Handbook of Intergenerational Ethics, New York, Oxford University Press, 2025*

Michel Bourban

Eco-Anxiety and Ecological Citizenship

Navigating an Ecological Emotion

Michel Bourban 🄳
University of Twente
Enschede, The Netherlands

ISBN 978-3-032-03218-8   ISBN 978-3-032-03219-5   (eBook)
https://doi.org/10.1007/978-3-032-03219-5

This work was supported by University of Twente.

Cover illustration: © Melisa Hasan

This Palgrave Macmillan imprint is published by the registered company Springer Nature Switzerland AG
The registered company address is: Gewerbestrasse 11, 6330 Cham, Switzerland

If disposing of this product, please recycle the paper.

For Lisa

PREFACE

Things are not looking good. When I began working on the philosophical problems raised by climate change some 15 years ago, the ambient mood was certainly not very cheerful, but there was at least a general feeling that climate change could be brought under political control in the foreseeable future. The IPCC had released its landmark fourth assessment report in 2007, unequivocally linking human activities to global warming and its planetary impacts, explaining in detail key mitigation measures for the short, medium, and long term, and providing a crucial input for international negotiations (IPCC 2007). Climate ethics was rapidly emerging as a recognized research field, with the first anthology published a few years later by the four leading voices in the field (Gardiner et al. 2010). It is true that the Copenhagen Accord, the outcome of COP15 in 2009, was disappointing (to say the least), as it contained no quantified commitments to mitigation targets, no binding mechanisms, and no verification measures. As a result, several countries did not sign this political agreement. However, the Cancun Agreements, which incorporated most of the outcomes of the Copenhagen conference, were almost unanimously adopted a year later. This new process was consolidated at successive meetings in Durban, Doha, Warsaw, and Lima. At the conclusion of COP20 in Lima, the common framework and guidelines for future negotiations were set out in the Lima Call for Action, which set the stage for a legally binding international agreement. This agreement was ultimately signed during COP21 in Paris in 2015, where the government representatives of the world committed to keep global warming "well below 2°C" and to "pursue efforts to limit the temperature increase to 1.5°C" (UNFCCC 2015, art. 2).

Over the past 15 years, the situation has deteriorated considerably. Far from being reduced or at least stabilized, global greenhouse gas (GHG) emissions have continued to increase. The average annual growth rate of GHG emissions was 2.1% between 2000 and 2009 and 1.3% between 2010 and 2019 (IPCC 2022). Global GHG emissions fell by 4.7% from 2019 to 2020 as a result of the lockdown measures taken to deal with the COVID-19 pandemic, but CO_2 emissions quickly returned to 2019 levels in 2021 (UNEP 2022). This trend continued in the following years, with an annual growth rate in GHG emissions of 1.2% in 2022 and 1.3% in 2023 and an increase of 0.9% in global CO_2 emissions in 2024 (Deng et al. 2025; UNEP 2023; 2024). As a result, the remaining global carbon budget has become very small and is rapidly shrinking. According to the most recent estimates, continued emissions at current levels would lead to global warming of 1.5°C in about five years' time (Forster et al. 2025). If all the mitigation pledges made by world governments are met, the world is on track to reach 2.8°C of global warming this century; if these (largely insufficient) pledges are not met, we are heading for 3.1°C (UNEP 2024).

In a landmark paper published in 2009 on the identification and quantification of planetary boundaries, Johan Rockström et al. (2009) stressed that to ensure a safe operating space for humanity, atmospheric concentration of carbon dioxide must not exceed 350 parts per million (ppm). In 2010, the global atmospheric CO_2 concentration was approximately 390 ppm; in early 2025, it reached 430 ppm (NASA 2025). This is a total increase of 40 ppm, or 2.7 ppm per year. As a result, the Earth system is currently on a "Hothouse Earth" trajectory, in which the living conditions of humans and countless other species are seriously threatened (Steffen et al. 2018). As Tim Lenton et al. (2019, 595) put it, "we are in a state of planetary emergency: both the risk and urgency of the situation are acute".

It is therefore no surprise that feelings of ecological worry, stress, and anxiety are on the rise. This is especially true, as we will see, for those who are already facing increasingly frequent and severe ecological impacts and also for those who are studying and carrying out research on environmental problems. Climate scientists report feeling high degrees of eco-anxiety. Children and young people are also among the most eco-anxious people. This has important implications for how educators approach environmental problems, as students also have strong—and legitimate—emotional responses to what and how we teach. Teaching environmental ethics and climate justice to undergraduate and graduate students for more than a decade has gradually led me to realize that the affective dimension of global environmental problems needs to be addressed explicitly in the classroom. We can no longer teach environmental studies as if this was an emotionally neutral topic: both teachers and students are emotionally affected by environmental problems. Even if most instructors (starting with me) are not really equipped with the tools to cope with this situation, more and more resources are available to deal with ecological emotions adequately in the classroom (see especially Atkinson and Ray 2024). This is an important pedagogical progress.

Our emotional lives can be quite complex and messy, and sharing and discussing our ecological emotions can be quite daunting, if not frightening, but avoiding the topic altogether is not a solution, as it leaves out of the discussion and reflection one of the most important impacts of environmental problems. I hope this book will be useful to researchers, educators, students, and other readers who, like me, are trying to navigate ecological emotions in the midst of the perfect ecological storm in which we find ourselves.

Enschede, The Netherlands   Michel Bourban

References

Atkinson, Jennifer, and Sarah Jaquette Ray, eds. 2024. *The Existential Toolkit for Climate Justice Educators: How to Teach in a Burning World*. 1st edn. University of California Press. https://doi.org/10.2307/jj.14284466.

Deng, Zhu, Biqing Zhu, Steven J. Davis, et al. 2025. 'Global Carbon Emissions and Decarbonization in 2024'. *Nature Reviews Earth & Environment* 6 (4): 231–33. https://doi.org/10.1038/s43017-025-00658-x.

Forster, Piers M., Chris Smith, Tristram Walsh, et al. 2025. 'Indicators of Global Climate Change 2024: Annual Update of Key Indicators of the State of the Climate System and Human Influence'. *Earth System Science Data* 17 (6): 2641–80. https://doi.org/10.5194/essd-17-2641-2025.

Gardiner, S. M., S. Caney, D. Jamieson, and H. Shue. 2010. *Climate Ethics: Essential Readings*. Oxford University Press.

IPCC. 2007. 'Summary for Policymakers'. In *Climate Change 2007: Synthesis Report. Contribution of Working Groups I, II and III to the Fourth Assessment Report of the Intergovernmental Panel on Climate Change*, edited by Rajendra Pachauri and Andy Reisinger. IPCC. https://www.ipcc.ch/site/ass ets/uploads/2018/02/ar4_syr_spm.pdf.

IPCC. 2022. 'Summary for Policymakers'. In *Climate Change 2022: Mitigation of Climate Change. Contribution of Working Group III to the Sixth Assessment Report of the Intergovernmental Panel on Climate Change*, edited by P. R. Shukla, J. Skea, R. Slade, et al. Cambridge University Press.

Lenton, Timothy M., Johan Rockström, Owen Gaffney, et al. 2019. 'Climate Tipping Points—Too Risky to Bet Against'. *Nature* 575 (7784): 592–95. https://doi.org/10.1038/d41586-019-03595-0.

NASA. 2025. *Carbon Dioxide*. https://climate.nasa.gov/vital-signs/carbon-dio xide/?intent=121.

Rockström, Johan, Will Steffen, Kevin Noone, et al. 2009. 'A Safe Operating Space for Humanity'. *Nature* 461 (7263): 472–75. https://doi.org/ 10.1038/461472a.

Steffen, Will, Johan Rockström, Katherine Richardson, et al. 2018. 'Trajectories of the Earth System in the Anthropocene'. *Proceedings of the National Academy of Sciences* 115 (33): 8252–59. https://doi.org/10.1073/pnas.181 0141115.

UNEP, ed. 2022. *The Closing Window: Climate Crisis Calls for Rapid Transformation of Societies*. The Emissions Gap Report 2022. United Nations Environment Programme.

UNEP. 2023. *Emissions Gap Report 2023: Broken Record—Temperatures Hit New Highs, yet World Fails to Cut Emissions (Again)*. United Nations Environment Programme. https://doi.org/10.59117/20.500.11822/43922.

UNEP. 2024. *Emissions Gap Report 2024: No More Hot Air … Please! With a Massive Gap between Rhetoric and Reality, Countries Draft New Climate Commitments*. United Nations Environment Programme. https://doi.org/ 10.59117/20.500.11822/46404.

UNFCCC. 2015. *Adoption of the Paris Agreement. Decision 1/CP.21. Document FCCC/CP/2015/10/Add.1*.

Acknowledgements I started working on eco-anxiety in 2019 for a presentation at a research seminar organized by Ulrich Hoinkes at Kiel University. Ulrich then generously invited me to transform this presentation into a chapter to be published in a book as part of the Anxiety Culture Project (Michel Bourban. 2024. "Eco-Anxiety: A Philosophical Approach", in Allegrante, John P., Hoinkes, Ulrich, Schapira, Michael I. and Struve, Karen (eds.), *Anxiety Culture: The New Global State of Human Affairs*, Baltimore: Johns Hopkins University Press: 55–69). I then wrote a second chapter on eco-anxiety for a handbook co-edited by Joel Kassiola on environmental politics and theory (Michel Bourban. 2023. "Eco-Anxiety and the Responses of Ecological Citizenship and Mindfulness", in Kassiola, Joel J. and Luke, Timothy W. (eds.), *The Palgrave Handbook of Environmental Politics and Theory*. Cham: Palgrave Macmillan: 65–88). I am very grateful to Ulrich for introducing me to the notions of anxiety culture and eco-anxiety and to Joel for his support and enthusiastic feedback on the chapter that led to this book project.

Over the last few years, I have been presenting parts of my research on eco-anxiety to different audiences that have helped me shape the ideas developed in this book. I would like to express my gratitude here to the participants in the Labont Annual Seminar organized by the Istituto Universitario di Studi Superiori Pavia and the Sant'Anna School of Advanced Studies (2021, online), the Environmental Philosophy of Technology (EPT) Workshop at the University of Twente (UT) (2023), the Living Room Philosophy Seminar organized by the Philosophy of Science and Technology in Society Study Association at UT (2023), the Southern African Society for Environmental Philosophy Inaugural Conference at Kruger National Park (2024), The European Philosophical Society for the Study of Emotions (EPSSE) Annual Conference at the University of Lisbon and Nova University of Lisbon (2024), the Ethics of Socially Disruptive Technologies (ESDiT) and 4TU.Ethics Annual Conference at UT (2024), the Climate Ethics at the Crossroads Conference at Sorbonne University Abu Dhabi (2024, online), the Environmental Humanities Colloquium at the University of Fribourg (2025, online), the ESDiT Away Days at Hotel De Keizerskroon in Apeldoorn (2025), the Anxiety and (Policy) Preference in Relation to Climate Change Workshop at Kiel University (2025), the More (Green) Energy to Combat (Natural) Disaster Blended Intensive Programme at Kiel University (2025), the

Behavioural, Management and Social Sciences (BMS) Research Conference at UT (2025), and the EPT Reading Group on the draft manuscript of the book at UT (2025). While this book draws on a wide range of contexts and conversations, some of the research underpinning it was carried out within the framework of the ESDiT programme (NWO grant number 024.004.031), and parts of it benefited from exchanges with colleagues and collaborative work within the Anxiety Culture Project.

For our discussions on eco-anxiety and/or ecological citizenship during the above-mentioned events and on other occasions, I am particularly thankful to John Allegrante, Pierre André, Bennet Francis, Linde Franken, Andrea Gammon, Stephen Gardiner, Jeroen Hopster, James Hutton, Elias König, Charlie Kurth, Angela Martin, Marjolein Oele, Eric Pommier, Alexandria Poole, Dominic Roser, Anne van Valkengoed, and Ivo Wallimann-Helmer. I would especially like to express my gratitude to Lisa Broussois and Dominic Lenzi, who both made detailed comments on the draft manuscript that helped me to substantially improve it. I would also like to thank the anonymous reviewer supplied by Palgrave Macmillan, who offered constructive feedback both on the book project and the draft manuscript, as well as Tania Doney, whose copyediting work greatly improved the text.

For their support and confidence in this project, I am grateful to Amy Invernizzi, Senior Editor in Philosophy at Palgrave Macmillan, and Antony Sami, Production Editor at Springer. I would also like to thank Madison Allums, formerly Commissioning Editor in US Politics and Political Theory at Palgrave Macmillan, who originally reached out to me to discuss the idea of writing a book on eco-anxiety.

Finally, I would like to thank Marit van Eck and Kasandra Poague, both Information Specialists for the Faculty of BMS at UT. Marit assisted me in the application process for the BMS Open Access Fund, which made it possible for this book to be published in open access format, and Kasandra designed the first two figures in Chapter 1.

Competing Interests The author has no competing interests to declare that are relevant to the content of this manuscript.

CONTENTS

LIST OF FIGURES

List of Tables

Introduction: Living Beyond Planetary Boundaries

Abstract This chapter contextualizes the topic of eco-anxiety by linking it to the anxiety culture phenomenon and the Anthropocene theoretical framework. The rise of eco-anxiety is directly related to the rise of anxiety levels over the last few decades and to long-lasting trends in the Earth system. The chapter explains the three main factors that make eco-anxiety a pervasive feature of life in the Anthropocene: the growing empirical data on the rapidly degrading state of the planet; the increasingly ubiquitous effects of global environmental changes; and the increasing number of fictional productions on ecological issues and their impacts on human and non-human beings. It also categorizes eco-anxiety as a threat-related ecological emotion and explains that even though it is not necessarily a new phenomenon, it increasingly affects people on a global scale and across different socio-demographic categories, albeit to different degrees.

Keywords Anxiety culture · Eco-anxiety · Climate anxiety · Anthropocene · Planetary boundaries · Ecological emotions

A RAPIDLY GROWING TOPIC OF RESEARCH

A few years ago, the notions of "eco-anxiety" and "climate anxiety" were still a marginal topic in academic research. As Panu Pihkala (2020, 1) stressed in a paper published in 2020, although both notions are much

© The Author(s) 2026

M. Bourban, *Eco-Anxiety and Ecological Citizenship*,
https://doi.org/10.1007/978-3-032-03219-5_1

discussed in newspaper articles, documentaries, blogs, and other media, "there is still a strong lack of research about various forms of such anxieties and about their relation to other psychological impacts of the ecological crisis".

Things have changed rapidly over the last few years. According to the Scopus database, a total of 965 academic publications are available on the topic as I write this in September 2025.[1] While only 36 academic studies were published on eco-anxiety or climate anxiety in 2020, in 2024 alone, 301 papers were published on the topic (see Fig. 1.1 – note that the data for 2025 is incomplete). This is also illustrated by the number of books that have been written on eco-anxiety and climate anxiety over the last few years, with a total of 25 books currently listed on the topic on Scopus (see, e.g., Corkett et al. 2025; Kennedy-Woodard and Kennedy-Williams 2022; Ray 2020; Schapira 2024; Vakoch and Mickey 2022; Wray 2022).

The most important disciplines in which eco-anxiety and climate anxiety have been investigated are social sciences (21.5% of the literature), psychology (16.5%), medicine (15.7%), and environmental science (13.2%) (see Fig. 1.2). Surprisingly, the topic is still relatively neglected in arts and humanities (7.2%), especially in philosophy. Another search in the same database reveals that only 57 publications are available on the topic from an explicitly philosophical or ethical perspective.[2] This is about 6% of the available literature. This can be explained by different factors, such as the newness of the topic (the proverbial owl of Minerva spreading its wings only as dusk falls[3]) or the fact that environmental philosophy and the (predominantly phenomenological) philosophical literature on anxiety have evolved mostly in parallel.

This book aims to bridge this gap in the literature in three complementary ways: (1) by proposing a detailed conceptual analysis of eco-anxiety, (2) by explaining how eco-anxiety raises ethical and justice-related issues,

---

[1] The search was conducted on 22 September 2025 on Scopus, based on the following search query: "TITLE-ABS-KEY ("eco-anxiety" OR "ecological anxiety" OR "climate anxiety")".

[2] The search was also conducted on 22 September 2025 on Scopus, based this time on the following search query: "TITLE-ABS-KEY ((("eco-anxiety" OR "ecological anxiety" OR "climate anxiety") AND (philosoph* OR ethic*)))".

[3] This famous phrase comes from the Preface to the *Philosophy of Right*, where Hegel writes, "the owl of Minerva begins its flight only with the onset of dusk". This proverb is often interpreted to mean that philosophy only comes in retrospect (Hegel 1991, 23).

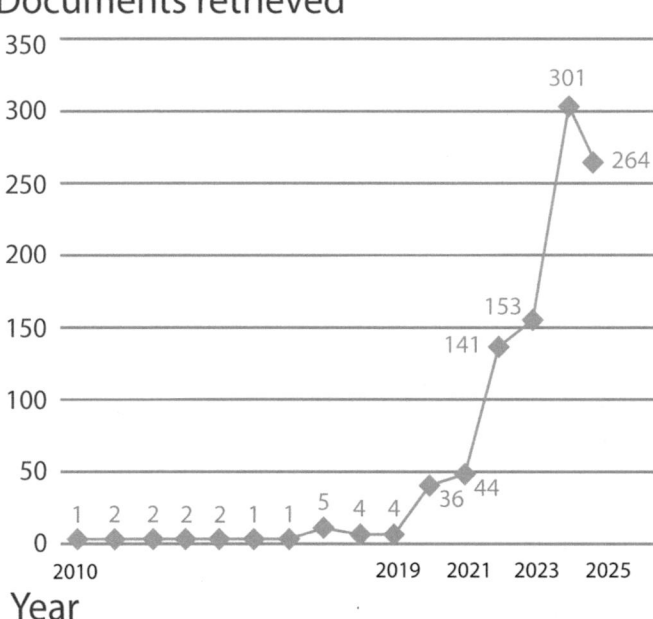

Fig. 1.1 The number of publications on eco-anxiety and climate anxiety by year (Credit: Kasandra Poague, based on data provided by Scopus)

and (3) by exploring a promising way to address these issues from an ecological citizenship perspective. To do so, it builds on both the slowly emerging philosophical literature on eco-anxiety and the more established literature in psychology and social sciences (on the emerging literature on eco-anxiety in philosophy, see Davidson 2024; Kurth and Pihkala 2022; Mosquera 2022; Oele 2024; Oksala 2023; Ott and Urner 2024; Pihkala 2022; Vakoch and Mickey 2022; Vaškovic 2023).

ANXIETY CULTURE AND THE ANTHROPOCENE

To understand why eco-anxiety has become such an important object of research in social and environmental sciences, it is important to place it in the context of the anxiety culture that characterizes the new global state of human affairs (Allegrante et al. 2024). Anxiety is not only a

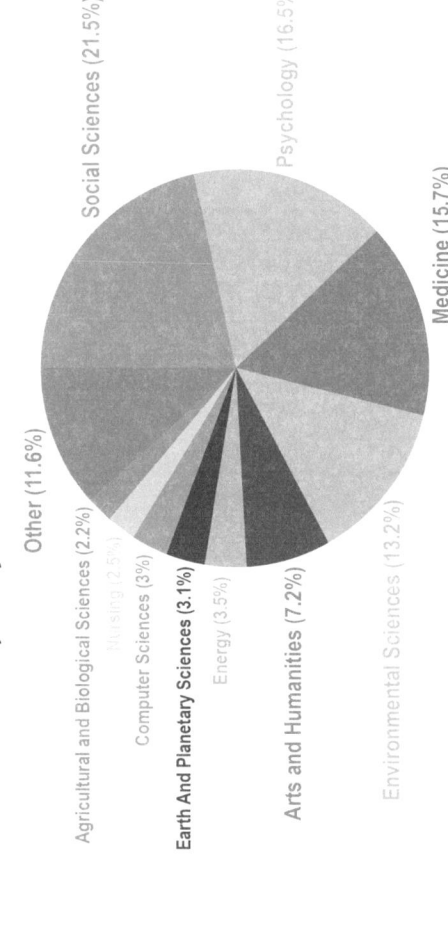

Fig. 1.2 The percentage of publications on eco-anxiety and climate anxiety by subject area (Credit: Kasandra Poague, based on data provided by Scopus)

clinical phenomenon; it is also a socio-cultural phenomenon that shapes the contours of human experience in response to pressing planetary challenges.

Anxiety culture is an intellectual and theoretical construct that helps to make sense of human life in contemporary societies. Multiple reasons make anxiety one of the defining features of our time: climate change and its increasingly frequent and severe impacts on populations and ecosystems, the increasing rate of species extinction, the global wave of autocratization and the erosion of democracies, wars and armed conflicts, the prospect of a nuclear war, the threat of terrorist attacks, the long-lasting effects of the COVID-19 pandemic, and AI applications and the prospect of artificial general intelligence that could surpass human abilities. All these challenges with ripple effects at the global level feed into growing feelings of uncertainty, insecurity, and powerlessness.

Anxiety is a complex affective state that manifests through our thoughts, feelings, and behaviours, and the next chapter will say much more about it. Suffice it to say for now that feelings of uncertainty, insecurity, and powerlessness are on the rise due to a cascade of anxiogenic crises that have characterized the last two decades through global environmental changes, the decline of democratic regimes and rise of autocratic ones, new wars and armed conflicts destabilizing the security of whole regions, the global threat of nuclear weapons, global pandemics, rapid technological changes, and rising socio-economic inequalities. Not only are there more and more anxiogenic factors, but today, we also have more access to detailed information on these factors. It is no coincidence that the rise of anxiety levels is taking place at the same time as a surge in the use of digital technologies such as the internet and social media. By digitalizing information and communication, these technologies are providing unprecedented access to knowledge related to these and other anxiogenic factors. The overabundance of digital content can also contribute to misinformation, disinformation, propaganda, and infowhelm, which can in turn feed feelings of anxiety.

Anxiety culture manifests itself in the connection between the state of the world and the global state of mental health. Young people are particularly affected, with unprecedented rates of anxiety disorders, depression, and suicide. Between 2009 and 2017, rates of depression among 18- to 25-year-old US residents rose 63%, with the number of young people experiencing serious psychological distress rising by 71% (Twenge et al. 2019). This phenomenon is pluri-causal. Unhealthy forms of digital communication play a major role. Climate change and other environmental problems also play an increasingly important role. This is highlighted by the 2020 WHO, UNICEF, and Lancet report, *A Future for the World's Children?*, which observes that children throughout the world are now facing existential threats from ecological degradation, the climate crisis, and predatory marketing of unhealthy and addictive products, with no country adequately protecting their health (Clark et al. 2020).

One of the major features of anxiety culture is our entry into the Anthropocene. The Anthropocene designates a new geological epoch in which some human activities rival the great forces of nature and threaten the living conditions and even the survival of multiple forms of life on the planet. It represents a rupture in Earth history characterized by the disruption of the Earth system and the (potentially irreversible) crossing of planetary boundaries (Hamilton 2016; Rockström et al. 2023; Steffen et al. 2015, 2018). The notion of "Anthropocene" is controversial for several reasons. First, it tends to be too general in its attribution of causal responsibility for global environmental problems and to neglect the fact that these problems are mainly caused by industrialized, capitalistic society, and this has led to proposals to replace "Anthropocene" with "Capitalocene" or "Anglocene", among other more precise alternatives (Bonneuil and Fressoz 2017). Second, in March 2024, the International Commission on Stratigraphy rejected the proposal of an Anthropocene Epoch as a formal unit for the Geologic Time Scale (Witze 2024). Third, even among those who agree with using this expression to designate a new epoch, there are disagreements regarding its starting date. For some, the Anthropocene started in the post-Second World War era, during the so-called Great Acceleration (Steffen et al. 2015); for others, it started with the Industrial Revolution and the invention of the steam engine (Crutzen and Stoermer 2000); for others still, it started with the agricultural revolution (Gowdy and Krall 2013).

I use the Anthropocene here as a general theoretical framework to refer to our time of profound and rapid changes in the Earth system. These changes are unprecedented in human history, both spatially and temporally. While, in the Holocene, humans' ecological footprint was always local and relatively short-lived, in the Anthropocene, the economic activities of industrialized and industrializing countries are altering basic natural life support systems on a global scale and for millennia. The Anthropocene is not only characterized by problems of pollution and environmental degradation that are limited spatially and temporally to the size and dynamics of biotic communities and ecosystems; it is first and foremost defined by global and intergenerational disruptions in the Earth system, such as climate change, biodiversity loss, and ocean acidification. This new kind of environmental degradation is best understood as a disruption in planetary-level systems, pushing the whole Earth system into a new state that is much less hospitable to humans and other species.

Beyond their global scale and intergenerational scope, Anthropocene environmental problems are also characterized by their irreversibility. According to the IPCC (2021, 6, 21), "The scale of recent changes across the climate system as a whole—and the present state of many aspects of the climate system—are unprecedented over many centuries to many thousands of years". They add, "Many changes due to past and future greenhouse gas emissions are irreversible for centuries to millennia". It is not always clear what the notion of "irreversibility" means in this context, as it is not explicitly conceptualized or defined in most of the scientific literature using this terminology. However, it usually refers to the type of system response relative to the forcing applied on it (abrupt, sudden), the quality (nonlinear, severe, rapid), and the timescale (long-term; centuries to millennia) of change in the biosphere, cryosphere, and hydrosphere, or the atmosphere and climate (Buhr et al. 2024). Whatever the quality and timescale of the environmental changes we are observing today, it is safe to say that socio-economic trends such as population growth, energy use, and water use contribute to long-lasting Earth system trends such as climate change, ocean acidification, and biodiversity loss, and that these trends are making our world a more dangerous place to live in.

The Anthropocene is characterized not only by the duration, severity, and irreversibility of environmental problems, but also by the awareness we have of the existence and severity of these problems. This awareness has developed gradually over the last decades, with several key reference points, such as Rachel Carson's *Silent Spring* in 1962 (Carson 2000), the Club of Rome report *Limits to Growth* in 1972 (Meadows et al. 1972), the Brundtland report *Our Common Future* in 1987 (WCED 1987), and the first IPCC assessment report in 1990 (Houghton et al. 1990), to mention just a few landmark publications. Our knowledge of the long-term risks caused by our economic activities and technical power is becoming increasingly substantial and accurate, to the extent that it is no longer possible (or at least excusable) to claim ignorance over the impacts these activities and this power have on the natural environment. This is why the Anthropocene and eco-anxiety go hand in hand.

PLANETARY BOUNDARIES, CLIMATE IMPACTS, AND ANTHROPOCENE FICTIONS

Eco-anxiety is a pervasive feature of life in the Anthropocene. This is due to three major factors: (1) the more and more numerous, accurate, and accessible empirical data on the rapidly degrading state of the planet; (2) the increasingly ubiquitous manifestations of rapid and severe ecological change; and (3) the increasing number of eco-fiction and climate-fiction narratives in films, television series, and novels. Let us discuss each in turn.

First, a large and growing part of the scientific community recognizes today that human activities are disrupting the Earth system by exceeding natural boundaries in planetary systems (Steffen et al. 2015, 2018; Richardson et al. 2023). There are nine planetary boundaries (see Fig. 1.3): the "big three" (the climate system, the ozone layer, and the ocean); the four biosphere boundaries (biodiversity, land, fresh water, and nutrients); and the two aliens (novel entities and aerosols). Six planetary systems have been pushed beyond their critical limits: the climate system, biosphere integrity (biodiversity), the nitrogen and phosphorus cycles (nutrients), land use, freshwater and green water, and novel entities (especially plastics). If these transgressions persist, the entire planet may be pushed into a new state that would be much less hospitable for human societies, not to mention other species.

Take the climate boundary. Crossing the planetary threshold of 2 °C above preindustrial temperature levels could lead the entire Earth system into a "Hothouse Earth" trajectory, in which global warming may be substantially accelerated. According to Will Steffen et al. (2018, 8254), "Currently, the Earth System is on a Hothouse Earth pathway driven by human emissions of greenhouse gases and biosphere degradation". This means that the Hothouse Earth trajectory is not just a possibility; it is also *more likely* than other trajectories in the Earth system, such as the Stabilized Earth pathway, in which global temperatures are maintained below 2 °C above the preindustrial level.

This is confirmed by the IPCC *Sixth Assessment Report* (AR6), which highlights that we are currently on a trajectory of global warming of 3.2 °C (with a range of 2.2 °C to 3.5 °C) by the end of the century if we maintain current policies and mitigation strategies (IPCC 2023, 22).

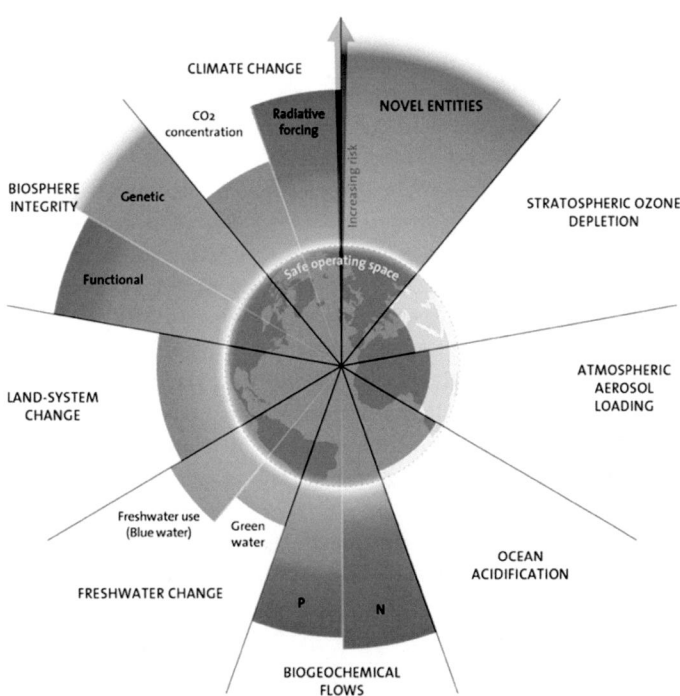

Fig. 1.3 The Nine Planetary Boundaries (2023 update). Licensed under CC BY-NC-ND 3.0 (Credit: Azote for Stockholm Resilience Centre, based on analysis in Richardson et al. 2023.)

Note that the "current policies and mitigation strategies" condition can cut both ways: mitigation policies can become more ambitious and lead us back to a less dangerous or even safe trajectory, but they can also weaken over time or be less effective than initially planned and lead to global warming that is even worse by the end of the century. Similarly, the "range of 2.2 °C to 3.5 °C" can also play in our favour or against us: the radiative forcing caused by increased greenhouse gas (GHG) emissions can prove to be lower or even higher than anticipated. Importantly, biogeophysical feedback processes within the Earth system may play a more important role than assumed in IPCC scenarios, which could trigger a Hothouse Earth pathway in which global temperature rise by 2100 could be even worse than that (Steffen et al. 2018).

A Hothouse Earth pathway would be triggered by a cascade of tipping points in the climate system. Tipping points are self-perpetuating changes beyond a warming threshold that lead to substantial and widespread impacts in the Earth system. The nine core tipping points are the two ice sheets in Greenland and West Antarctica, the Arctic and East Antarctic Sea ice, the Boreal Forest and the Amazon rainforest, Atlantic circulation, permafrost in Siberia, and Australia's Great Barrier Reef. Each tipping point has a respective warming threshold, but David Armstrong McKay et al. (2022) have found that current global warming (about 1.1 °C above the preindustrial level) lies within the lower end of five climate tipping points. This means that we are already at risk of triggering tipping points. Six tipping points become likely within the Paris Agreement range (1.5 °C to < 2 °C), including the collapse of the Greenland and West Antarctic ice sheets, the die-off of low latitude coral reefs, and widespread and abrupt permafrost thaw (see Fig. 1.4). As Tim Lenton et al. (2019, 595) add,

> the evidence from tipping points alone suggests that we are in a state of planetary emergency: both the risk and urgency of the situation are acute […]. The intervention time left to prevent tipping could already have shrunk towards zero, whereas the reaction time to achieve net zero emissions is 30 years at best.

This brings us to the second factor that explains the pervasiveness of eco-anxiety: the fact that multiple consequences of global environmental changes are already being observed and experienced today and are becoming increasingly severe. Even if we manage to avoid a Hothouse Earth pathway, there will still be plenty of adverse environmental impacts on human and non-human beings. This is also confirmed by the IPCC AR6, which lists observed impacts of climate change both on ecosystems, with changes in ecosystem structure, species range shifts, and changes in timing, and on human systems, with impacts on water scarcity and food production, on health and well-being, and on cities, settlements, and infrastructures (IPCC 2022).

Widespread and pervasive impacts on ecosystems and human systems have, for instance, been caused by more frequent and intense climate and weather extremes, such as hot extremes on land and in the ocean, heavy

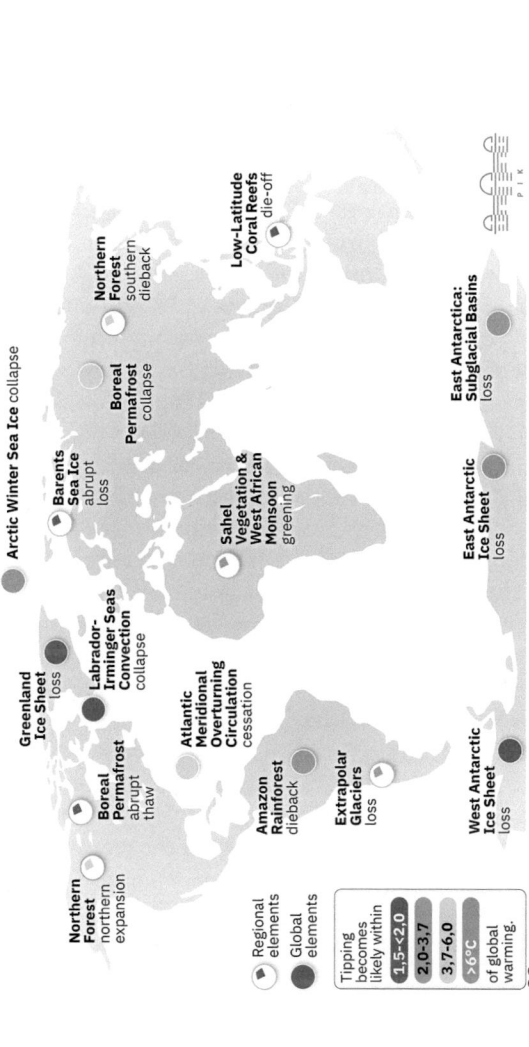

Fig. 1.4 The geographical distribution of global and regional tipping elements, colour-coded according to the best estimate for their temperature thresholds, beyond which the element would be likely to be "Tipped". Figure designed at PIK (under CC BY licence)

precipitation, and drought and fire events, which led to increased heat-related human mortality, coral bleaching and mortality,[4] and drought-related tree mortality. Climate change has already caused substantial damages and losses, with widespread deterioration of ecosystem structure and function, reduction in natural adaptive capacity, and shift in seasonal timing, which has led to hundreds of local irreversible losses of species and mass mortality events on land and in the ocean. It has also exposed millions of people to acute food insecurity and reduced water security, thereby hindering efforts to meet the Sustainable Development Goals (SDGs). The most disproportionally affected people and communities are living in Africa, Asia, Central and South America, small islands, and the Arctic, with approximately *half of the global human population* already experiencing severe water scarcity for at least some part of the year. Climate change is contributing to humanitarian crises, with climate and weather extremes increasingly causing forced human displacement in all regions, with Small Island Developing States (SIDS) being the most vulnerable. Approximately *3.3 to 3.6 billion people* live in contexts that are disproportionately affected by climate change and its impacts (IPCC 2022, 9–12).

Importantly, both the changes we are currently observing in the Earth system and their impacts on human and non-human beings are long-lasting. Take climate change. Part of the past, present, and future CO_2 emissions will be absorbed by the biosphere and the oceans, but another part will remain in the atmosphere almost forever. The time it takes for the chemical composition of the atmosphere to adjust to emissions of carbon dioxide can be divided into three phases associated with increasing timescales (Ciais et al. 2014, 472–473). At the end of the first phase, which lasts 1000 years, the remaining amount of CO_2 emitted into the atmosphere is between 15 and 40%, depending on the amount of carbon released by human activities. The rest will have been absorbed by the biosphere and the oceans, but the higher the initial emissions, the less carbon those two sinks will be able to absorb. The second phase, which takes about 10,000 years, reduces the remaining fraction of atmospheric

---

[4] In a report published while I was finishing writing this book, Timothy Lenton et al. (2025) warn that warm-water coral reefs are already crossing their thermal tipping point and experiencing unprecedented dieback. This is the first tipping point that has been crossed. Parts of the polar ice sheets may also have crossed tipping points, which could eventually lead to an irreversible sea-level rise of several metres.

CO_2 to between 10 and 25% of the original emissions. It is only in the final phase, which lasts several hundred thousand years, that the rest of the carbon dioxide emitted by humans will be removed from the atmosphere. Past and present emissions of CO_2 will, therefore, remain in the atmosphere for *tens or even hundreds of thousands of years*, contributing to climate impacts during all that time. And the more of this GHG we emit, the higher the percentage remaining in the atmosphere will be.

One major implication of this is that eco-anxiety is here to stay. This is not only because global environmental changes will continue to develop and worsen over time, but also because the risks they create for human and non-human beings will become more severe and more likely over time. In many scenarios, ecological problems and the risks they pose can be mitigated, but (1) things will in all likelihood continue to worsen before they improve because of the current socio-economic and Earth system trends, and (2) many of the political and technological measures taken so far to address these problems have proven largely insufficient or even inadequate, as illustrated by the repeated failures of international negotiations to bring global GHG emissions under control. Multiple mitigation and adaptation measures are within reach, but many depend on collective action and coordination and will therefore be difficult to adopt and implement, especially at a time where multilateralism is threatened and already undermined by the rise of nationalism throughout the world and the global wave of autocratization.[5] As we will see in Chapter 3, this is not a reason to give in to eco-despair, but it is a strong reason to feel eco-anxious.

The third factor that explains why eco-anxiety is so widespread is the increased production of fictional narratives on global environmental changes and the effects of these changes on human and non-human beings. More and more Anthropocene fictions are exploring apocalyptic and/or post-apocalyptic scenarios in which anthropogenic ecological disasters completely change the world as we know it (Johns-Putra 2016; Svoboda 2016; Trexler 2015; Trexler and Johns-Putra 2011). There are multiple films, television series, documentaries, and novels portraying

---

[5] In another recent report, the V-Dem Institute highlights that for the first time in over 20 years, there are now fewer democracies than autocracies, with liberal democracies becoming the least common political regime in the world. As a result, nearly three out of four people in the world (that is, 72% of the world population) now live in autocracies (Nord et al. 2025).

possible ecological disasters in the nearer or more distant future that might trigger eco-anxiety.[6] It is not necessary to directly experience environmental impacts or disasters in order to feel eco-anxious; imagining environmental risks and dangers can be sufficient.

Anthropocene fictions link numbers, graphs, and figures with our daily lives. They bring the notion of the Anthropocene to life and make it tangible. They represent possible interpretations of the notion of ecological disaster, allowing the viewer or reader to apprehend this notion through imagination and emotions rather than through reason alone. Portrayals of ecological disaster are particularly frequent in the (post-) eco-apocalyptic subgenre, which traces the end of the world as we know it back to environmental problems that are happening now or to the potentially catastrophic side effects of some of the technologies we might use to address them.

Eco-Anxiety as an Ecological Emotion

Living in a state of planetary emergency, beyond planetary boundaries and in the shadow of tipping points, can cause a range of existential, psychological, and emotional responses. Global environmental changes such as biodiversity loss, climate change, and ocean acidification have a strong affective dimension. These affective phenomena include mental states such as feelings, moods, and emotions.

Glenn Albrecht (2019, ix) defines Earth emotions as the "particular human emotional responses we have to the scale and pace of ecological and environmental change". Such emotions are tied to planetary states that can be both felt and perceived; they refer to the relationship between the biophysical environment and our mental health. One major ecological emotion discussed by Albrecht is solastalgia, which he defines as "the homesickness you have when you are still at home. […] Home is becoming more than unrecognizable: it is for many becoming increasingly hostile" (Albrecht 2019, 200). Solastalgia is an example of *sadness-related*

---

[6] Examples include documentaries such as Jeff Orlowski's *Chasing Coral*, David Attenborough and Johan Rockström's *Breaking Boundaries*, and Fisher Stevens and Leonardo DiCaprio's *Before the Flood*; eco-fiction movies such as Kevin Reynolds' *Waterworld*, Roland Emmerich's *The Day After Tomorrow*, and Bong Joon-ho's *Snowpiercer*; and climate-fiction novels such as Ian McEwan's *Solar*, Ronald Wright's *A Scientific Romance*, and Kim Stanley Robinson's *Science in the Capital* trilogy.

ecological emotion, a category of Earth emotions that also includes environmental melancholia, "a condition in which even those who care deeply about the well-being of ecosystems and future generations are paralyzed to translate such concern into action" (Lertzman 2015, 4) and ecological grief, or eco-grief, "the grief felt in relation to experienced or anticipated ecological losses, including the loss of species, ecosystems, and meaningful landscapes" (Cunsolo and Ellis 2018, 275).

Sadness-related ecological emotions can be contrasted with *threat-related ecological emotions* (Pihkala 2022). Such emotions include for instance Anthropocene horror, "a sense of horror about the changing environment globally, usually as mediated by news reports and expert predictions, giving a sense of threats that need to be anchored to any particular place, but which are both everywhere and anywhere" (Clark 2020), as well as global dread: "the anticipation of an apocalyptic future state of the world that produces a mixture of terror and sadness in the sufferer for those who will exist in such a state" (Albrecht 2019, 199). As this last example shows, there is a certain degree of overlap between sadness-related and threat-related ecological emotions; however, this taxonomy of ecological emotion, which can be further developed to include for instance self-condemning as well as other-condemning emotions, is useful for making sense of the diversity of emotional states tied to global environmental changes.

Eco-anxiety and climate anxiety belong to the category of threat-related ecological emotions. While eco-anxiety can be broadly defined as a "chronic fear of environmental doom" (Clayton et al. 2017, 29), climate anxiety has been characterized as a "persistent, difficult-to-control apprehensiveness and worry about climate change" (Van Valkengoed et al. 2023, 258). They are both related to fear, apprehensiveness, and worry. The difference between the two notions is that while climate anxiety refers to a form of eco-anxiety focused on climate change and its future impacts, eco-anxiety also covers other environmental changes and their consequences, such as biodiversity loss, ocean acidification, and freshwater change. Eco-anxiety is the most generic term, which is why I use it the most in this book, even though many examples I take of environmental impacts focus on climate change and therefore are also relevant to climate anxiety. In other words, climate anxiety is a form of eco-anxiety that is often discussed in the rest of the book, but the approach I adopt is broader and includes both climate risks and other ecological risks.

Although climate anxiety is a relatively new phenomenon caused by the multiplication of climate impacts throughout the world and the development of the scientific study of climate change through climatology, atmospheric science, oceanography, Earth system science, and so on, eco-anxiety is not necessarily new. It might seem like a new psychological state for middle-class citizens of industrialized and rich countries who are starting to feel increasingly threatened by global environmental changes, but individuals and groups disproportionately affected by environmental impacts are all too familiar with this kind of ecological emotion (Wray 2022, 3). For instance, Australian Aboriginal people, like many Indigenous people throughout the world, have a double experience of turmoil, with colonial and climatic forces leading to persistent economic damage and cultural loss (Albrecht 2019, x). As Kyle Whyte (2017) points out, climate change is an intensification of environmental change imposed on Indigenous people by historical injustices related to colonialism, capitalism, and industrialization.[7] Current climate impacts are very similar to the effects Indigenous people had to endure because of colonialism: ecosystem degradation, species loss, economic crash, relocation, and cultural disintegration (Whyte 2018). This is part of the phenomenon of emotional injustice discussed in Chapter 2: those who are the least responsible for climate and environmental impacts are disproportionately affected by them, both physically and psychologically. Although eco-anxiety is not necessarily a new phenomenon, it is definitely becoming a much more widespread ecological emotion.

Importantly, eco-anxiety is not separated from other kinds of ecological emotions. This is especially the case for eco-grief, which is triggered by experienced or anticipated ecological losses. Just like grief, eco-grief involves a series of affective states, such as denial, anger, joy, and anxiety. Eco-anxiety can be felt especially in the case of projected or anticipated loss of species, ecosystems, and landscapes. This connection between eco-grief and eco-anxiety is stressed by Ashlee Cunsolo and Neville Ellis (2018, 278), when they stress that anticipatory grief is "grief emergent from anxiety of, or preparation for, future losses and mourning for an anticipated future that will likely cease to be". They have observed this

---

[7] This is why the development of Indigenous studies based on Indigenous knowledges is a crucial area of research and can contribute, as Whyte (2017, 153) puts it, to "indigenizing futures" by "decolonizing the Anthropocene".

mix of eco-grief and eco-anxiety associated with anticipated future ecological losses in both Inuit communities in Northern Canada and farming communities in rural Australia, where many individuals are grieving over future losses to culture, livelihoods, and ways of life. It is true that while eco-anxiety is a threat-related, forward-looking emotion, eco-grief is more a sadness-related, backward-looking emotion. However, they can overlap when the experience or anticipation of ecological loss leads to a feeling of insecurity related to an uncertain future. Eco-grief can actually be both backward- and forward-looking: one can feel grief about an actual loss and/or an anticipated loss, and both losses can lead to feelings of eco-anxiety. The relationship between these two ecological emotions is therefore a dynamic one.

A Global and Socially Widespread Phenomenon

How many people actually are eco-anxious? The data available to answer this question is still limited, but there is an emerging empirical literature that provides a general picture of the state of eco-anxiety throughout the world and society, with a strong focus on children and young people.

A review of surveys at the national level highlights that while 40% of British people aged 16–24 describe feeling overwhelmed because of the environmental emergency, 57% of American teenagers report that climate change makes them feel scared. In Australia, 96% of people aged 7–25 consider climate change to be a serious problem, with 89% saying they are worried about the effects of climate change, and 70% being concerned that adults do not or will not take their opinion on climate change seriously (Kennedy-Woodard and Kennedy-Williams 2022).

The largest study to date on eco-anxiety surveyed 10,000 children and young people aged 16–25 in ten countries: Australia, Brazil, Finland, France, India, Nigeria, Philippines, Portugal, the UK, and the US (Hickman et al. 2021). It found that 59% of children and young people report that they feel very or extremely worried about climate change, with 45% saying that their feelings about climate change negatively affect their daily lives (see Fig. 1.5). More specifically, 83% think that people have failed to take care of the planet, 75% find the future frightening, 56% believe that humanity is doomed, 55% think that the things they most value will be destroyed, 55% say that they will have fewer opportunities than their parents, 52% think that their family security will be threatened, and 39% are hesitant to have children. In total, more than 50% say that they feel afraid, sad, anxious, angry, powerless, helpless, and/or guilty.[8] Factors such as location and age play an important role in whether, and to what extent, people are eco-anxious, but empirical evidence shows that experiences of ecological anxiety are rapidly increasing (Cunsolo et al. 2020; Clayton et al. 2017, 2021). This also explains why the literature

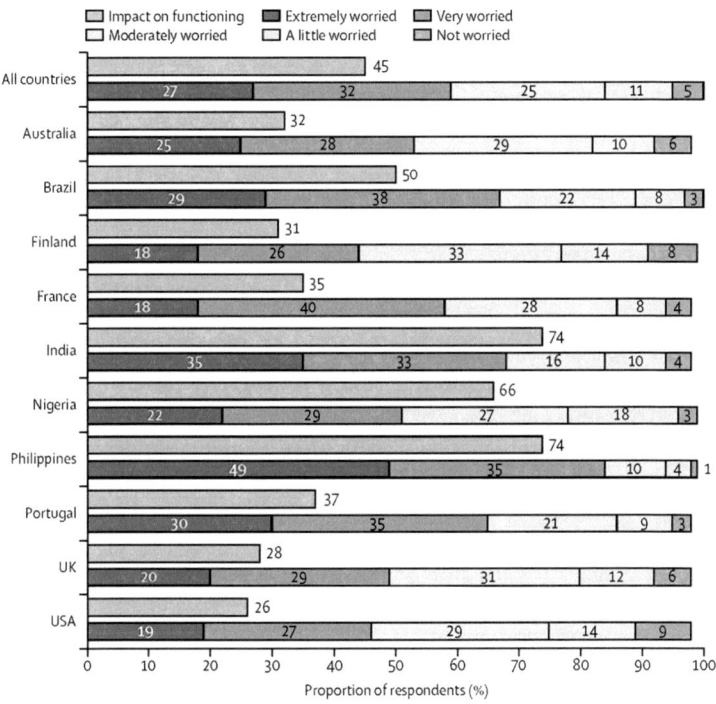

Fig. 1.5 Worry about climate change and impact on functioning. Licensed under CC BY-NC-ND 4.0 (Credit: Hickman et al. (2021))

---

[8] These numbers raise an important methodological question: how is eco-anxiety measured? Although there are different options to measure it (see, e.g., Clayton and Karazsia 2020), one influential method is the Hogg Eco-Anxiety Scale, a psychological measurement tool assessing eco-anxiety levels on the basis of a 13-item scale capturing four dimensions of eco-anxiety: affective symptoms, rumination, behavioural symptoms, and anxiety about one's negative impact on the planet (Hogg et al. 2021).

on the topic has developed so rapidly in recent years. Together with other ecological emotions such as solastalgia, environmental melancholia, ecological grief, Anthropocene horror, and global dread, eco-anxiety has become a pervasive characteristic of life in the Anthropocene.

Even if eco-anxiety is becoming increasingly widespread, there are two points that should be stressed: first, there are different degrees of eco-anxiety, and second, not everyone is feeling eco-anxious. As we will see in Chapter 2, one factor that can make eco-anxiety so difficult to live with is that many people around eco-anxious people tend to downplay the relevance or even challenge the appropriateness of this ecological emotion. After all, as Fig. 1.5 shows, only 26% of respondents in the US, 28% in the UK, 31% in Finland, 32% in Australia, and 35% in France said their feelings about climate change are negatively affecting their daily lives (Hickman et al. 2021).

Still, these numbers need to be nuanced by two other findings. The first is that in many other countries, the numbers are much higher, with 50% of respondents in Brazil, 66% in Nigeria, 74% in the Philippines, and 74% in India reporting that their feelings about climate change are negatively affecting their lives on a daily basis. The second is that when we look at how children and young people perceive their future, the geographical differences are much weaker, with 75% of respondents worldwide and 73% in the UK who think that the future is frightening and over half who think that humanity is doomed (56% worldwide and 51% in the UK). As Hickman (2024, 358) stresses, "It is when we look at how they see their futures, how they navigate the world today, what hope they have for the future that we see how powerfully children are being affected even in countries that are not facing the worst impact of climate change, yet".

Instead of looking at a specific age category worldwide, it is also possible to focus on a specific country. Another recent study reviews the state of eco-anxiety in France, surveying 998 people aged 15–64 (Sutter et al. 2025). At first sight, it might seem that this report found that a small minority of the French population is eco-anxious (see Table 1.1). By extrapolating their findings to the totality of the French population based on predictive analysis, the authors explain that while 75% of the total population have little or no eco-anxiety, only 15% are moderately eco-anxious, and 10% more strongly eco-anxious. While this seems to imply that eco-anxiety is a secondary or peripheral problem in France, a closer look at the data indicates otherwise.

Table 1.1   Estimated number of eco-anxious people in France, by intensity category

Intensity category	Percentage of population	Extrapolation: number of people concerned
Not at all eco-anxious	5.7%	2.4 million
Very slightly eco-anxious	44.3%	18.6 million
Slightly eco-anxious	25%	10.5 million
Moderately eco-anxious	15%	6.3 million
Strongly eco-anxious	5%	2.1 million
Very strongly eco-anxious ...	5%	2.1 million
... at risk of depression or anxiety disorder	1%	420,000

Adapted from Sutter, Bernaud, and Messmer (2025, 11)

First, if we look at absolute rather than relative numbers, the study shows that 6.3 million people are moderately eco-anxious with initial symptoms that must be kept in check, 2.1 million are severely eco-anxious, and another 2.1 million are very severely eco-anxious, to the point that they require psychological treatment. This is more than *10 million people*. In the category of very severely eco-anxious people, 420,000 are at risk of depression or anxiety disorder. Second, even if we stick to relative numbers, the study found that only 5.7% are not at all eco-anxious; the largest share of the population (44.3%) reports being very slightly eco-anxious, followed by those who report being slightly eco-anxious (25%). Even though these last two population categories have fairly limited levels of eco-anxiety and therefore are not dealing with significant mental health challenges related to environmental problems, they still feel somewhat eco-anxious, which means that this ecological emotion is more widespread than it might seem at first. This represents more than *19 million people*. These two reasons explain why the report concludes that "eco-anxiety can be considered a public health issue" (Sutter et al. 2025, 11).

Another relevant finding of this report is that "eco-anxiety affects French people of all ages, contrary to the common misconception that eco-anxiety only affects young people" (Sutter et al. 2025, 12). The study finds that younger people, aged 15–24, are not the most eco-anxious in France; it is the 25–34 age group that has the highest level of eco-anxiety, with the 50–64 age group displaying the lowest levels of eco-anxiety. This finding also needs to be nuanced.

First, the 15–24 age group follows the 25–34 age group closely in terms of intensity of eco-anxiety, stressing that "the two youngest age categories (15–24 and 25–34) are the most exposed to the greatest threats to mental health" (Sutter et al. 2025, 12). Second, the global survey by Hickman et al. (2021) found that in the 16–25 age group in France, 40% said they felt very worried about climate change and 18% felt extremely worried (cf. Fig. 1.4). These results contrast with those from the national survey by Sutter et al. (2025), where only 5% of the 15–24 age group were strongly eco-anxious and 8% very strongly eco-anxious. These contrasting results reflect differences in methodological approaches and underline the importance of interpreting these numbers with caution, but they do suggest eco-anxiety levels among children and young people in France are relatively high.

A final important finding worth mentioning at this stage is that, in addition to age group, other socio-demographic categories can act as factors that are predictive of eco-anxiety levels. One key factor is gender, with women being significantly more likely than men to be eco-anxious as they are in the majority in the groups of moderately (56%), strongly (63%), and very strongly (54%) eco-anxious people in France (Sutter et al. 2025). One reason for this might be that women tend to be more anxious in general because of biological and psychosocial factors (Farhane-Medina et al. 2022). Another reason could be that women also tend to be more environmentally aware than men. This is illustrated by the so-called green gender gap, which reveals that in the US, 62% of climate voters are women and that young women are twice as likely as young men to list climate change as their top priority (Environmental Voter Project 2025). Other key determining factors include the area where someone lives, with people living in the Paris region being on average more eco-anxious than in other regions of France, and the size of the conurbation, with people living in large conurbations being on average more eco-anxious than people living in rural communities. Other factors also play a more minor role, such as professional activity, with farmers being on average twice as eco-anxious as pensioners, and education level, with people with no academic qualifications or lower levels of educational qualification appearing to be less eco-anxious than those with higher levels of education. All this indicates that eco-anxiety is a socially widespread phenomenon. People from more than one socio-demographic group disproportionately exposed to eco-anxiety are the most at risk, for instance young, highly educated women or children living in large conurbations.

These predictive factors have also been discussed in other studies. In their review of the literature on the health implications of eco-anxiety, Inmaculada Boluda-Verdú et al. (2022, 1) found that there is a "link between eco-anxiety and negative mental health outcomes, mainly in younger generations, women, and poorer countries in the 'Global South'". Those with pre-existing health conditions, with limited resources, and whose lives and health are directly threatened by severe climate impacts, are also more likely to be more eco-anxious. Clayton et al. (2023) also found that gender, age, and country play an important role in psychological and emotional responses to climate change.[9] They also found that women are not only more at risk of being negatively impacted by climate change and more environmentally aware and concerned, but also that they tend to be more eco-anxious than men, who tend to be more optimistic and express greater faith in governmental action. They also stress that children and young people in the Philippines, India, and Nigeria have to deal with a stronger psychological impact of climate change than those in the US, the UK, and Finland. In these countries, younger generations are bound to experience higher impact on functioning because they are directly threatened by some of the most severe climate impacts.

STRUCTURE OF THE BOOK

After these introductory considerations, the rest of the book will develop my account of eco-anxiety, which follows two main objectives: to specify and develop a definition of eco-anxiety through a conceptual clarification analysis; and to link eco-anxiety with ecological citizenship to investigate a promising but underexplored way to address eco-anxiety.

Chapter 2 builds a definition of eco-anxiety based on the literature on anxiety and eco-anxiety, with the objective of moving from a vague and short definition to a longer, more specific account of eco-anxiety, stressing both its negative and positive aspects. It shows that the building blocks of eco-anxiety are future orientation, a focus on ecological risks

---

[9] These results were found using the dataset collected by Hickman et al. (2021).

and dangers, environmental awareness and insecurity, and manifestations such as risk-assessment and risk-minimization behaviours (in the case of low and moderate forms of eco-anxiety), as well as feelings of discouragement and paralysis (in the case of more severe forms of eco-anxiety). This chapter also explains that even if eco-anxiety is not a mental disorder, it is important to clearly distinguish between different degrees of eco-anxiety to identify its most severe forms and its more worrying manifestations. Importantly, it also stresses that there is something inherently rational about eco-anxiety, as it usually represents, at least at first, a lucid reaction to the deteriorating state of our natural environment. The tension between the certainty of the existence and severity of ecological risks and the uncertainty raised by their development in the future is a core feature of eco-anxiety.

Chapter 3 takes the understanding of eco-anxiety developed in Chapter 2 and looks at promising ways to cope with it. As eco-anxiety is a pervasive ecological emotion in the Anthropocene, getting rid of it is neither possible, nor, as we will see, desirable; it is therefore important to find how to learn to live with this ecological emotion. Ecological citizenship is presented as a form of moral and political community that helps with addressing some of the most worrying effects of eco-anxiety, such as paralysis and isolation, and that can contribute to harnessing some of the positive effects of eco-anxiety, such as a motivation to undertake environmental actions, both at the individual and collective levels. This chapter also discusses the role of hope in facing eco-anxiety, stressing that this emotion can be a double-edged sword when it comes to motivating eco-anxious people to undertake environmental action, and explaining why good hope and radical hope are the most relevant forms of hope in this context. Anxiety and hope are complex emotions, which have both positive and negative sides. Just as it does not make much sense to call eco-anxiety a "negative" emotion, it is not necessarily the case that hope represents a "positive" emotion. It all depends on which aspect of these complex emotions we focus on. Finally, this chapter proposes an environmental virtue ethics approach focused on carbon sobriety and courage, which are conceived as green virtues helping the ecological citizen to live with eco-anxiety and combat paralysis and despair.

The conclusion sums up the main findings of the analysis and suggests areas of future research to further develop the account presented here. It provides an answer to four of the most important questions of the book: what is eco-anxiety? Why is it a moral and political problem? What is new about this emotion? And what can be done to deal with it? By doing so, it points to possible ways to learn to live with this ecological emotion.

REFERENCES

Albrecht, Glenn. 2019. *Earth Emotions: New Words for a New World*. Cornell University Press.

Allegrante, John P., Ulrich Hoinkes, Michael I. Schapira, and Karen Struve, eds. 2024. *Anxiety Culture: The New Global State of Human Affairs*. Johns Hopkins University Press.

Armstrong McKay, David I., Arie Staal, Jesse F. Abrams, et al. 2022. 'Exceeding 1.5°C Global Warming Could Trigger Multiple Climate Tipping Points'. *Science* 377 (6611): eabn7950. https://doi.org/10.1126/science.abn7950.

Boluda-Verdú, Inmaculada, Marina Senent-Valero, Mariola Casas-Escolano, Alicia Matijasevich, and María Pastor-Valero. 2022. 'Fear for the Future: Eco-Anxiety and Health Implications, a Systematic Review'. *Journal of Environmental Psychology* 84 (December): 101904. https://doi.org/10.1016/j.jenvp.2022.101904

Bonneuil, Christophe, and Jean-Baptiste Fressoz. 2017. *The Shock of the Anthropocene: The Earth, History and Us*. Verso.

Buhr, Lorina, Dominic S Lenzi, Auke J K Pols, et al. 2024. 'The Concepts of Irreversibility and Reversibility in Research on Anthropogenic Environmental Changes'. *PNAS Nexus* 4 (1): 577. https://doi.org/10.1093/pnasnexus/pgae577.

Carson, Rachel. 2000. *Silent Spring*. Reprinted. Penguin Books, Modern Classics. Penguin.

Ciais, P., C. Sabine, and W. Peters. 2014. 'Carbon and Other Biogeochemical Cycles'. In *Climate Change 2013—The Physical Science Basis*, 1st edn, edited by IPCC, claytonhogg. Cambridge University Press. https://doi.org/10.1017/CBO9781107415324.015.

Clark, Helen, Awa Marie Coll-Seck, Anshu Banerjee, et al. 2020. 'A Future for the World's Children? A WHO–UNICEF–Lancet Commission'. *The Lancet* 395 (10224): 605–658. https://doi.org/10.1016/S0140-6736(19)32540-1.

Clark, Timothy. 2020. 'Ecological Grief and Anthropocene Horror'. *American Imago* 77 (1): 61–80. https://doi.org/10.1353/aim.2020.0003.

Clayton, Susan, and Bryan T. Karazsia. 2020. 'Development and Validation of a Measure of Climate Change Anxiety'. *Journal of Environmental Psychology* 69 (June): 101434. https://doi.org/10.1016/j.jenvp.2020.101434.

Clayton, Susan, Christie Manning, Kirra Krygsman, and Meighen Speiser. 2017. *Mental Health and Our Changing Climate: Impacts, Implications, and Guidance*. American Psychological Association & ecoAmerica.

Clayton, Susan, Christie Manning, Meighen Speiser, and Alison N. Hill. 2021. *Mental Health and Our Changing Climate: Impacts, Inequities, Responses.* American Psychological Association & ecoAmerica.

Clayton, Susan, Panu Pihkala, Britt Wray, and Elizabeth Marks. 2023. 'Psychological and Emotional Responses to Climate Change among Young People Worldwide: Differences Associated with Gender, Age, and Country'. *Sustainability* 15 (4): 3540. https://doi.org/10.3390/su15043540.

Corkett, Julie K., Wafaa Mohammed Moawad Abd-El-Aal, and Astrid Steele, eds. 2025. *Addressing Climate Anxiety in Schools: Pedagogical Perspectives and Theoretical Foundations.* 1st ed. Research and Teaching in Environmental Studies. Routledge.

Crutzen, Paul J., and Eugene F. Stoermer. 2000. 'The « Anthropocene »'. *Global Change Newsletter* 41: 17–18.

Cunsolo, Ashlee, and Neville R. Ellis. 2018. 'Ecological Grief as a Mental Health Response to Climate Change-Related Loss'. *Nature Climate Change* 8 (4): 275–281. https://doi.org/10.1038/s41558-018-0092-2.

Cunsolo, Ashlee, Sherilee L. Harper, Kelton Minor, Katie Hayes, Kimberly G. Williams, and Courtney Howard. 2020. 'Ecological Grief and Anxiety: The Start of a Healthy Response to Climate Change?' *The Lancet Planetary Health* 4 (7): e261–e263. https://doi.org/10.1016/S2542-5196(20)30144-3.

Davidson, Joe Pl. 2024. 'The Politics of Eco-Anxiety: Anthropocene Dread from Depoliticisation to Repoliticisation'. *The Anthropocene Review* 11 (2): 427–41. https://doi.org/10.1177/20530196231211854.

Environmental Voter Project. 2025. *The Green Gender Gap: How Women Are Shaping the Climate Vote.* Environmental Voter Project. https://www.enviro nmentalvoter.org/sites/default/files/documents/green-gender-gap-report. pdf.

Farhane-Medina, Naima Z., Bárbara. Luque, Carmen Tabernero, and Rosario Castillo-Mayén. 2022. 'Factors Associated with Gender and Sex Differences in Anxiety Prevalence and Comorbidity: A Systematic Review'. *Science Progress* 105 (4): 00368504221135469. https://doi.org/10.1177/003685042211 35469.

Gowdy, John, and Lisi Krall. 2013. 'The Ultrasocial Origin of the Anthropocene'. *Ecological Economics* 95 (November): 137–147. https://doi.org/10.1016/j. ecolecon.2013.08.006.

Hamilton, Clive. 2016. 'The Anthropocene as Rupture'. *The Anthropocene Review* 3 (2): 93–106. https://doi.org/10.1177/2053019616634741.

Hegel, Georg Wilhelm Fredrich. 1991. *Hegel: Elements of the Philosophy of Right:* 1st edn. Edited by Allen W. Wood. Translated by H. B. Nisbet. Cambridge University Press. https://doi.org/10.1017/CBO9780511808012.

Hickman, Caroline. 2024. 'Eco-Anxiety in Children and Young People—A Rational Response, Irreconcilable Despair, or Both?' *The Psychoanalytic Study of the Child* 77 (1): 356–368. https://doi.org/10.1080/00797308.2023. 2287381.

Hickman, Caroline, Elizabeth Marks, Panu Pihkala, et al. 2021. 'Climate Anxiety in Children and Young People and Their Beliefs about Government Responses to Climate Change: A Global Survey'. *The Lancet Planetary Health* 5 (12): e863–e873. https://doi.org/10.1016/S2542-5196(21)00278-3.

Hogg, Teaghan L., Samantha K. Stanley, Léan. V. O'Brien, Marc S. Wilson, and Clare R. Watsford. 2021. 'The Hogg Eco-Anxiety Scale: Development and Validation of a Multidimensional Scale'. *Global Environmental Change* 71 (November): 102391. https://doi.org/10.1016/j.gloenvcha.2021.102391.

Houghton, John, Geoffrey Jenkins, and John Ephraums, eds. 1990. *Climate Change: The IPCC Scientific Assessment: Report Prepared for Intergovernmental Panel on Climate Change by Working Group I*. Cambridge University Press.

IPCC. 2021. 'Summary for Policymakers'. In *Climate Change 2021: The Physical Science Basis. Contribution of Working Group I to the Sixth Assessment Report of the Intergovernmental Panel on Climate Change*, edited by V. Masson-Delmotte, P. Zhai, A. Pirani, et al. Cambridge University Press.

IPCC. 2022. 'Summary for Policymakers'. In *Climate Change 2022: Impacts, Adaptation, and Vulnerability. Contribution of Working Group II to the Sixth Assessment Report of the Intergovernmental Panel on Climate Change*, edited by H.-O. Pörtner, D. C. Roberts, E. S. Poloczanska, et al. Cambridge University Press.

IPCC. 2023. 'Summary for Policymakers'. In *Synthesis Report of the IPCC Sixth Assessment Report*, edited by H. Lee, K. Calvin, D. Dasgupta, et al. Cambridge University Press.

Johns-Putra, Adeline. 2016. 'Climate Change in Literature and Literary Studies: From Cli-Fi, Climate Change Theater and Ecopoetry to Ecocriticism and Climate Change Criticism'. *Wires Climate Change* 7 (2): 266–282. https://doi.org/10.1002/wcc.385.

Kennedy-Woodard, Megan, and Patrick Kennedy-Williams. 2022. *Turn the Tide on Climate Anxiety: Sustainable Action for Your Mental Health and the Planet*. With Arizona Muse. Jessica Kingsley Publishers.

Kurth, Charlie, and Panu Pihkala. 2022. 'Eco-Anxiety: What It Is and Why It Matters'. *Frontiers in Psychology* 13 (September): 981814. https://doi.org/10.3389/fpsyg.2022.981814.

Lenton, Timothy M., Manjana Milkoreit, Simon Willcock, et al. 2025. *The Global Tipping Points Report 2025*. University of Exeter. https://global-tipping-points.org.

Lenton, Timothy M., Johan Rockström, Owen Gaffney, et al. 2019. 'Climate Tipping Points—Too Risky to Bet Against'. *Nature* 575 (7784): 592–595. https://doi.org/10.1038/d41586-019-03595-0.

Lertzman, Renee. 2015. *Environmental Melancholia*. 0 edn. Routledge. https://doi.org/10.4324/9781315851853.

Meadows, Donella, Dennis L. Meadows, Jergen Randers, and William Behrens III. 1972. *The Limits to Growth*. Universe Books.

Mosquera, Julia. 2022. 'How to Feel about Climate Change? An Analysis of the Normativity of Climate Emotions'. *International Journal of Philosophical Studies* 30 (3): 357–80. phl.

Nord, Marina, David Altman, Fabio Angiolillo, Tiago Fernandes, Ana Good God, and Staffan I. Lindberg. 2025. *Democracy Report 2025: 25 Years of Autocratization—Democracy Trumped?* V-Dem Institute. https://v-dem.net/documents/61/v-dem-dr__2025_lowres_v2.pdf.

Oele, Marjolein. 2024. 'Anxiety, Grief, and Trust in Times of Climate Change: A Phenomenology of Affective Constellations and Future Transformations in and beyond the Anthropocene'. *Comparative and Continental Philosophy*, June 3, 1–20. https://doi.org/10.1080/17570638.2024.2361409.

Oksala, Johanna. 2023. 'The Existential Threat of Climate Change: From Climate Anxiety to Post-Nihilist Politics'. *Environmental Philosophy* 20 (2): 191–214. https://doi.org/10.5840/envirophil2023919133.

Ott, Konrad, and Maren Urner. 2024. 'Climate Anxieties in Discourse: From Mental States to Arguments'. In *Anxiety Culture: The New Global State of Human Affairs*, edited by John P. Allegrante, Ulrich Hoinkes, Michael I. Schapira, and Karen Struve. Johns Hopkins University Press.

Pihkala, Panu. 2020. 'Anxiety and the Ecological Crisis: An Analysis of Eco-Anxiety and Climate Anxiety'. *Sustainability* 12 (19): 7836. https://doi.org/10.3390/su12197836.

Pihkala, Panu. 2022. 'Toward a Taxonomy of Climate Emotions'. *Frontiers in Climate* 3 (January): 738154. https://doi.org/10.3389/fclim.2021.738154.

Ray, Sarah Jaquette. 2020. *A Field Guide to Climate Anxiety: How to Keep Your Cool on a Warming Planet*. University of California Press.

Richardson, Katherine, Will Steffen, Wolfgang Lucht, et al. 2023. 'Earth beyond Six of Nine Planetary Boundaries'. *Science Advances* 9 (37): eadh2458. https://doi.org/10.1126/sciadv.adh2458.

Rockström, Johan, Joyeeta Gupta, Dahe Qin, et al. 2023. 'Safe and Just Earth System Boundaries'. *Nature* 619 (7968): 102–111. https://doi.org/10.1038/s41586-023-06083-8.

Schapira, Kate. 2024. *Lessons from the Climate Anxiety Counseling Booth: How to Live with Care and Purpose in an Endangered World*. First edition. Hachette Go Books.

Steffen, Will, Wendy Broadgate, Lisa Deutsch, Owen Gaffney, and Cornelia Ludwig. 2015. 'The Trajectory of the Anthropocene: The Great Acceleration'. *The Anthropocene Review* 2 (1): 81–98. https://doi.org/10.1177/2053019614564785.

Steffen, Will, Johan Rockström, Katherine Richardson, et al. 2018. 'Trajectories of the Earth System in the Anthropocene'. *Proceedings of the National Academy of Sciences* 115 (33): 8252–8259. https://doi.org/10.1073/pnas. 1810141115.

Sutter, Pierre-Eric, Jean-Luc Bernaud, and Léonie Messmer. 2025. *Éco-Anxiété En France (Étude 2025)*. ADEME. https://librairie.ademe.fr/societe-et-politi ques-publiques/8137-eco-anxiete-en-france.html.

Svoboda, Michael. 2016. 'Cli-Fi on the Screen(s): Patterns in the Representations of Climate Change in Fictional Films'. *Wires Climate Change* 7 (1): 43–64. https://doi.org/10.1002/wcc.381.

Trexler, Adam. 2015. *Anthropocene Fictions: The Novel in a Time of Climate Change*. University of Virginia Press.

Trexler, Adam, and Adeline Johns-Putra. 2011. 'Climate Change in Literature and Literary Criticism'. *Wiley Interdisciplinary Reviews: Climate Change* 2 (2): 185–200. https://doi.org/10.1002/wcc.105.

Twenge, Jean M., A. Bell Cooper, Thomas E. Joiner, Mary E. Duffy, and Sarah G. Binau. 2019. 'Age, Period, and Cohort Trends in Mood Disorder Indicators and Suicide-Related Outcomes in a Nationally Representative Dataset, 2005–2017'. *Journal of Abnormal Psychology* 128 (3): 185–199. https://doi. org/10.1037/abn0000410.

Vakoch, Douglas A., and Sam Mickey, eds. 2022. *Eco-Anxiety and Planetary Hope: Experiencing the Twin Disasters of COVID-19 and Climate Change*. Springer International Publishing. https://doi.org/10.1007/978-3-031-084 31-7.

Valkengoed, Van, M. Anne, Linda Steg, and Peter De Jonge. 2023. 'Climate Anxiety: A Research Agenda Inspired by Emotion Research'. *Emotion Review* 15 (4): 258–262. https://doi.org/10.1177/17540739231193752.

Vaškovic, Petr. 2023. 'Philosophical Perspectives on Climate Anxiety'. In *Handbook of the Philosophy of Climate Change*, edited by Gianfranco Pellegrino and Marcello Di Paola. Handbooks in Philosophy. Springer International Publishing. https://doi.org/10.1007/978-3-031-07002-0_144.

WCED. 1987. *Our Common Future*. Oxford University Press.

Whyte, Kyle. 2017. 'Indigenous Climate Change Studies: Indigenizing Futures, Decolonizing the Anthropocene'. *English Language Notes* 55 (1–2): 153–162. https://doi.org/10.1215/00138282-55.1-2.153.

Whyte, Kyle P. 2018. 'Indigenous Science (Fiction) for the Anthropocene: Ancestral Dystopias and Fantasies of Climate Change Crises'. *Environment and Planning E: Nature and Space* 1 (1–2): 224–242. https://doi.org/10.1177/ 2514848618777621.

Witze, Alexandra. 2024. 'It's Final: The Anthropocene Is Not an Epoch, despite Protest over Vote'. *Nature*, March 20, d41586-024-00868-1. https://doi. org/10.1038/d41586-024-00868-1.

Wray, Britt. 2022. *Generation Dread: Finding Purpose in an Age of Climate Crisis*. Alfred A. Knopf Canada.

Eco-Anxiety: Understanding an Ecological Emotion

Abstract This chapter defines eco-anxiety through a detailed analysis of its major conceptual building blocks. The definition focuses on ecological risks as the object of eco-anxiety, on environmental awareness and the related feeling of insecurity, and on behavioural responses ranging from risk-assessment and risk-minimization to discouragement and paralysis. The chapter shows that eco-anxiety revolves around a core tension between certainty and uncertainty. It also explains that, even though eco-anxiety should not be categorized as a medical condition, some forms of eco-anxiety can lead to mental disorders. To help separate healthy from unhealthy forms of eco-anxiety, the chapter builds a distinction between different degrees of eco-anxiety. Finally, the chapter explains that eco-anxiety has crucial moral and political implications, as it can lead to moral injury, the violation of the human right to health, and emotional injustice.

Keywords Anxiety · Eco-anxiety · Mental health · Risk · Technology · Moral injury · Human rights · Emotional injustice

This chapter does not systematically review the literature on eco-anxiety or climate anxiety; several other publications have already done so (e.g. Boluda-Verdú et al. 2022; Coffey et al. 2021; Cosh et al. 2024; Gago et al. 2024; Ojala et al. 2021). The main objective here is to identify the key features of eco-anxiety, building on the literature available on the

© The Author(s) 2026   31
M. Bourban, *Eco-Anxiety and Ecological Citizenship*,
https://doi.org/10.1007/978-3-032-03219-5_2

topic, but complementing it with a more detailed definition and more developed conceptual analysis. The point is to get a better idea of what eco-anxiety is, which will then serve to think about possible ways to cope with it in the next chapter. As stressed in Chapter 1, both eco-anxiety and climate anxiety are relevant to the analysis, but I focus on the former as it represents the most generic category and covers a larger set of ecological risks.

This chapter begins by drawing on the philosophical and psychological literature on anxiety to start identifying some key features of eco-anxiety. Next, it provides a working definition of eco-anxiety focused on three key elements: ecological risks as the object of eco-anxiety; the role of environmental awareness and lucidity; the variety of behavioural responses. It also explains why eco-anxiety is not a medical condition, even if severe and very severe forms of eco-anxiety can lead to pathological responses, and why it is not only a psychological problem but also a moral and political one. The main objective is to address what I shall call the conceptual vagueness and conceptual fragmentation problems in the literature through a detailed conceptual analysis of the notion of "eco-anxiety".

Anxiety: Some Basic Considerations

The notion of "eco-anxiety" is difficult to define. One major reason for this is that anxiety is a multifaced affective state that has always been difficult to grasp. Another reason is that scholars from many different disciplines are contributing to the literature on eco-anxiety (see Figure 1. 2), each with their own terminology, methodology, and approaches.

As Anne van Valkengoed et al. (2023, 258) highlight, "the conceptualisation of climate anxiety differs considerably across studies, which hinders gaining a clear understanding of what climate anxiety is, and comparing and integrating findings". Some studies use emotional responses as an indicator of eco-anxiety, such as nervousness, fear, and distress; others focus more on cognitive and behavioural manifestations, such as fatigue, sleep deprivation, and disruption of daily functioning. More critically, Charlie Kurth and Panu Pihkala (2022, 1) found that much of the existing work on eco-anxiety "has been hampered by conceptual and methodological difficulties". This is typically due to heterogeneous amalgamations between a wide range of ecologically oriented affective experiences such as pro-environmental beliefs and behaviours. This is also sometimes due to over-simplistic models reducing eco-anxiety to a mere threat-avoidance response. There is also a tendency to conflate eco-anxiety with other affective states, such as guilt and grief. Studies that systematically review the literature on eco-anxiety confirm that there is both a lack of agreement on the definition of eco-anxiety (call this the *conceptual fragmentation* problem) and a lack of conceptual clarity (call this the *conceptual vagueness* problem). While Hailie Brophy et al. (2023, 635) observe that there is a "lack of definition consensus of the concept", Inmaculada Boluda-Verdú et al. (2022, 1) explain that "A wide variety of eco-anxiety definitions was used in the different studies but further research is needed to provide conceptual clarity of the term eco-anxiety".

Because of the conceptual fragmentation and conceptual vagueness problems, there are persistent misunderstandings around the terms "eco-anxiety" and "climate anxiety", with a growing tendency to attribute a negative value to these affective states as sections of the public see the term or the state it refers to as stigmatizing (Gregersen et al. 2024). For this reason, Thea Gregersen et al. (2024, 6) suggest referring less to the term "anxiety" when discussing emotional and psychological responses to climate change and other environmental problems, and replacing it with expressions such as "climate worry" or "climate concern". I disagree. The problem with this suggestion that we should talk less about eco-anxiety is that it would maintain and even potentially strengthen the confusion around the notion and the stigmatization of eco-anxious people. I find it more relevant to keep using the term while dissipating current misunderstandings and misconceptions around it through a detailed conceptual analysis of what eco-anxiety is.

A good place to start looking for a definition of eco-anxiety is to examine the literature on anxiety. The notion of anxiety is even more complex than that of eco-anxiety. One reason for this is that anxiety corresponds to a variety of phenomena, such as social worries, hardwired responses to potential threats, existential angst, and clinical disorders (Kurth 2018b). Another reason is that there is a very rich tradition in modern and contemporary philosophy on anxiety from Kant to Levinas by way of Schelling, Kierkegaard, Darwin, Nietzsche, Freud, Husserl, and Heidegger (for a philosophical history of the notion, see Bergo 2021).[1] There is also no single definition of anxiety, although the one provided

---

[1] For contributions to eco-anxiety drawing on this philosophical tradition, see, e.g., Lafontaine (2022), Mickey (2022), Oele (2024), Vaškovic (2023), and Vaškovic and Vičanová (2024).

by the American Psychiatric Association (APA) in the *Diagnostic and Statistical Manual of Mental Disorders* (DSM) is often used as a starting point for discussions on the topic. According to the APA, anxiety may be defined as an "apprehensive anticipation of future danger or misfortune accompanied by a feeling of dysphoria or somatic symptoms of tension" (APA 2000, 820). In other words, anxiety is a future-oriented state in which one anticipates a danger with uneasy and unhappy feelings and physical manifestations of tension. It is first and foremost a state of uncertainty through which an event or situation is evaluated as implying an uncertain threat (Miceli and Castelfranchi 2005).

What kind of affective state does anxiety correspond to? The APA links anxiety with the notion of "feeling", which corresponds to an immediate state of mind. Importantly, the APA does not consider that anxiety *is* a feeling, but that it is "accompanied by a feeling of dysphoria". This feeling of uneasiness and unhappiness corresponds to what it feels like to be anxious, but it does not imply that anxiety is (only) a feeling. Anxiety is more often categorized as an *emotion* (Miceli and Castelfranchi 2005; Freeman and Freeman 2012; Kurth 2018b; Vazard and Kurth 2022). Even if there is little consensus on the nature of emotions, it is still helpful to categorize anxiety as an emotion, and more specifically as a bio-cognitive emotion based on two mechanisms: the hardwired, biological response to potential threats and the more flexible cognitive response to address the uncertainty at hand through behaviours such as risk-minimization, risk-assessment, deliberation, reflection, and information gathering (Kurth 2018b, 7–8).

More specifically, anxiety is an emotional response to uncertainty about what to do in the face of a threat or challenge whose potential is unpredictable, uncontrollable, or open to question. It is both a defensive response directed towards protecting oneself or someone else against physical harm or social threat and an epistemic response concerned with good or accurate decision making (Kurth 2018a). Three elements are important here: (1) the lack of information on the part of the subject; (2) the fact that this state of uncertainty is problematic as it is related to a possible challenge, threat, or danger; and (3) the fact that the subject recognizes this possible threat or challenge and tries to respond to it by engaging in epistemic behaviours that can help them decide on what is the right thing to do (Kurth 2018b, 1–6; Vazard and Kurth 2022). The third element is the main reason why Kurth (2018b, 131–136) considers that, at least in its moderate manifestations, anxiety is a valuable emotion: it helps the anxious person to recognize a range of threats and dangers and to respond to them. Anxiety brings a distinctive sensitivity and responsiveness to problematic uncertainty, it is a manifestation of the recognition of life's complexity and uncertainty. It makes us aware of the need to engage in reflection and deliberation and gives us a motivational push to do something about the threats and dangers that are constitutive of the uncertainty we are facing.

Two important points need to be stressed to complete this account, the first on uncertainty and the second on certainty. First, there is no guarantee that the anxious person will succeed in finding adequate ways to address the threat they have identified. In many cases, the anxious person does not know how to respond to the threat, even after having gathered relevant information on its causes. This can increase the state of uncertainty, which in turn leads to feeling even more anxious: the threat is identified, the relevant information has been found, but there is still no clear path of action that presents itself as a solution. This is a very anxiogenic situation, which, as we will see, is also relevant in the case of eco-anxiety.

Second, not everything is uncertain for the anxious person. The starting point of anxiety is often the *certainty* that a threat will materialize in the future, even if the likelihood, the severity, or the timing of the threat remains uncertain. In other words: the subject lacks information on the potential of the threat, which makes it unpredictable and uncontrollable, but at the same time, the subject knows that the threat exists and needs to be addressed. Take, for instance, the case that has been at the centre of discussions in the phenomenological literature on anxiety: the anxiety triggered by one's own mortality. In that case, the object of anxiety is very certain as there is no escaping one's own finitude. The same can be said about other reasons to feel anxious, such as the presence of pain, suffering, and other people's deaths in human existence, which are also all unavoidable, and represent what David Benatar (2017) calls the human predicament. However, we do not know how or when such events will occur. As we will see, this tension between certainty and uncertainty is also a constitutive feature of eco-anxiety.

Before turning to eco-anxiety, I would like to make two additional remarks. First, anxiety has also been categorized as another kind of affective state: a *mood*. While emotions are often considered as short-term reactions to a specific object, moods are considered as more pervasive affective states with no specific object (Kurth 2018b; Oele 2024; Vaškovic

2023; Vazard and Kurth 2022). Just like emotions, moods come with bodily sensations and feelings, but they usually last for longer. In the philosophical literature, it is the mood state of anxiety that has been the most important object of research. Having an anxious mood means having a long-lasting experience of a sense of worry that is not directed at anything clearly identifiable. Below, I follow the characterization of anxiety as an emotional response to problematic (un)certainty, but I also discuss one feature of eco-anxiety that is usually attributed to the mood of anxiety: the long-lasting sense of dread about what the future holds. I also briefly comment on the possibility of interpreting eco-anxiety as a mood.

Second, anxiety is not only one of the most fundamental human emotions; it is also one of the most common forms of *psychological disorder*. It is linked to a series of clinical problems, such as phobias, panic disorder, generalized anxiety disorder (GAD), and post-traumatic stress disorder (PTSD). About one-third of the adult population reports having anxiety problems, with almost one-fifth meeting the criteria for clinical disorder (Freeman and Freeman 2012, 111). It is therefore crucial not to underestimate the detrimental effects that anxiety can have on one's mental health. At the same time, it is also important not to reduce anxiety to a mental disorder—anxiety disorders are not anxiety tout court. Usually categorized as a negative emotion, anxiety tends to be presented as an inherently detrimental or intrinsically debilitating state. However, as stressed above, anxiety can also be a valuable affective state (Kurth 2018a, b; Vazard and Kurth 2022). By making potential threats salient, it allows us to prepare for the emergence of possible risks and dangers. While extreme forms of anxiety can indeed be detrimental and even impairing, more moderate forms can have beneficial outcomes, typically through enhancing our physical and intellectual performance and making us aware of all kinds of threats by pushing us to assess them and to address their causes. There are therefore clinical and non-clinical forms of anxiety, and it is an emotion with detrimental and beneficial sides to it. This will be important to keep in mind when defining eco-anxiety, a task to which I am turning now.

ECO-ANXIETY: A WORKING DEFINITION

Towards a More Specific Definition

How do these considerations help us to reflect on the main features of eco-anxiety? In an early contribution to research on ecological emotions, Glenn Albrecht (2011, 49) defined eco-anxiety as an "anxiety related to a changing and uncertain environment", showing the importance of uncertainty in eco-anxiety and linking it to environmental changes. Surprisingly, however, his more recent book *Earth Emotions* does not have much more to say about this key ecological emotion, which is not even mentioned in the glossary; it simply repeats the same definition, adding that "With future uncertainty being one of the hallmarks of climate change prediction, a generalized worry about the future is now commonplace" (Albrecht 2019, 76–77). Albrecht stresses that eco-anxiety now features in many academic and non-academic publications, but he does not engage with this literature. We, therefore, must look elsewhere to find additional conceptual elements.

A definition that is influential in the literature is the one provided by ecoAmerica and the APA in their important report *Mental Health and our Changing Climate*, in which eco-anxiety is conceived as a "chronic fear of environmental doom" (Clayton et al. 2017, 29). This definition does not provide much more information than Albrecht's, but it shows (1) that the kind of threat eco-anxious people must deal with is specifically environmental in nature, (2) that this state is persistent or constantly recurring, and (3) that it is a form of fear.

In the same vein, van Valkengoed et al. (2023, 258) define climate anxiety as a "persistent, difficult-to-control apprehensiveness and worry about climate change". There is also the idea of a long-lasting state focused on an environmental problem (in this case climate change) with the addition of the difficulty of controlling this state of apprehensiveness and worry. They also stress that climate anxiety comes with emotional (such as fear and nervousness), physiological (such as nausea and muscle tension), cognitive (fatigue, difficulty concentrating), and behavioural (restlessness, impairment in daily functioning) indicators. Also focusing on the behavioural component, Kurth and Pihkala (2022, 11) define eco-anxiety more positively as "an emotion that sensitizes us to difficult decisions about things like climate change, and that brings behaviours (reflection and engagement) aimed at helping us address the difficulty we face".

These definitions provide a good starting point to conceptualize eco-anxiety, but they remain too vague to precisely define it. This is probably intentional, as eco-anxiety is a complex phenomenon and trying to capture its major features in one definition might prove to be reductive. Vagueness has its virtues, especially when it comes to multifaceted emotions. However, I find it helpful to build on these existing efforts to provide a more detailed definition and capture more clearly the contours of the notion. By doing so, I neither wish to provide a definitive definition of eco-anxiety nor to excessively restrict our understanding of anxiety. It is in that spirit that I propose the following working definition[2]:

> Eco-anxiety is an ecological emotion turned towards a possible state of the planet or the more direct environment in the future. Its object is ecological risks that are not yet here (in space) or present (in time), but which might happen at some point in a nearer or more distant future. It is an emotional response to an acute awareness of real ecological risks that leads to a state of insecurity. In its more moderate forms, eco-anxiety can give rise to constructive risk-assessment and risk-minimization behaviours aimed at addressing ecological risks. In its more severe forms, it can lead to an overwhelming feeling of discouragement that can become paralysing.

This definition draws the contours of the core elements of eco-anxiety; at its periphery, eco-anxiety can be many other things, and this is where it becomes more difficult to distinguish it from other ecological emotions, such as eco-grief.

There are three main components to this definition: (1) the orientation towards an uncertain and threatening future; (2) the role of awareness of risks which leads to a state of insecurity; (3) the variety of behavioural responses to eco-anxiety, ranging from risk-assessment and risk-minimization to discouragement and paralysis. Let us discuss each in turn.

---

[2] This is a working definition because I keep revising it, and it will need to be further improved in future research. Earlier versions of this definition can be found in Bourban (2023) and Bourban (2024). The account that follows in this section is a developed and updated analysis that draws on these two previous contributions.

An Uncertain and Threatening Future

Like any form of anxiety, eco-anxiety is future-oriented. The envisioned future can be more or less determinate and nearer or further away, but it is both (a) uncertain and (b) threatening.

Regarding (a), risks are the product of magnitude and probability: the magnitude is a measure of the seriousness of the loss and damage at stake; the probability is a measure of the likelihood of this loss and damage occurring (Shue 2010). That something is uncertain means that it currently has no measurable probability; it does not mean that its objective probability, if known, would be small. Questions of probability can, however, become less relevant if the magnitude of the possible loss and damage is massive. Take the two core planetary boundaries through which all the other boundaries operate: the climate system and biosphere integrity. As mentioned in Chapter 1, a cascade of tipping points in the climate system, such as rapid permafrost thawing, weakening of terrestrial and oceanic carbon sinks, and Amazon forest dieback could lead the entire Earth system onto a Hothouse Earth pathway whereby global warming may be substantially accelerated (Steffen et al. 2018). Likewise, the sixth mass species extinction leads to the degradation of ecosystem services and jeopardizes the integrity of ecosystems, posing an existential threat to life support systems that are essential for both human and non-human beings (Ceballos et al. 2020). Due to the magnitude of the loss and damage at hand, accelerated climate disruption and accelerated biodiversity loss are more than mere risks; they represent "transcendental damages", that is, damages that threaten the very condition of human existence on the planet, or at least the conditions of a flourishing human life (Bourg 2013).

Regarding (b), eco-anxiety is not about any risk, but more specifically about *ecological risks*. Such risks include sea level rise, megafires, and extreme climate events such as hurricanes, droughts, heatwaves, and floods. The object of eco-anxiety can also be the effect of these phenomena on humans and human societies, such as the disruption of agricultural systems and food supply chains, economic shocks, and sociopolitical instability along with starvation, mass migration, conflict, and even societal collapse. As Johanna Oksala (2023, 201) stresses, eco-anxiety involves the "fear of the loss of a collective future" as well as the fear of "civilizational devastation". Similarly, Petr Vaškovic (2023, 473) links eco-anxiety with apocalyptic anxiety that is focused on the "future end of human civilization itself". However, the object of eco-anxiety is even wider than this; it also extends to the consequences of environmental changes on non-human beings, such as animal suffering, ecosystem degradation, and species extinction. This is why, as Marjolein Oele (2024, 7) rightly highlights, eco-anxiety "entangles multiple species" and is also oriented towards "more-than-human loss". This implies that eco-anxious people are not only or even mainly anxious about their own mortality, as in traditional philosophical accounts of anxiety, but also about the wellbeing of future people and non-human beings such as sentient animals, ecosystems, and other species. In other words, eco-anxiety has both anthropocentric and non-anthropocentric elements.

Importantly, ecological risks do not only include environmental problems and the negative impacts they might have on humans and nonhuman nature; they extend to the risks raised by measures aimed at addressing environmental problems. Take, for instance, the social and environmental impacts that energy transition policies can have if they are not designed and implemented in a just way. Regarding social impacts, renewable energy technologies can give rise to negative impacts on local communities, human rights abuses in global supply chains, exploitative North–South relations and green colonialism, and the exacerbation of injustices related to income and gender (Andreucci et al. 2023; Levenda et al. 2021; Murphy and Elimä 2021; Sovacool et al. 2019). Regarding environmental impacts, renewable energy technologies can negatively affect biodiversity and ecosystems as a result of increased mining, land use change, and water, soil, and air pollution (Gasparatos et al. 2017; Sonter et al. 2020).

Some of the technologies that are currently being discussed to address climate change can contribute to techno-anxiety, a specific form of anxiety that is primarily driven by uncertainties associated with the risks that come with technological innovation. This is typically the case with socially disruptive technologies (SDTs), that is, technologies that prevent important aspects of human society from continuing without (major) change, by challenging existing institutions, individual and social behaviours, and legal norms. SDTs can be anxiogenic because they raise significant uncertainty about future social, economic, and political development. As Jeroen Hopster (2021, 7) explains, SDTs represent "technologies that have deep, important, ethically salient and wide-ranging impacts, that occur rapidly, provoke uncertainty and cannot be easily reversed".

Take, for instance, the case of so-called geoengineering or "climate engineering" technologies, a broad category that refers to "*deliberate large-scale manipulation of the planetary environment to counteract anthropogenic climate change*" (Royal Society 2009, 1 – emphasis original). Climate engineering measures include solar radiation management (SRM) techniques, which aim to reduce solar radiation that reaches Earth. SRM methods such as stratospheric sulphur injection hold the potential to reduce surface air temperature, thereby reducing or reversing some of the negative impacts of global warming. However, some of the disruptions they may cause include significant risks, such as the creation of droughts in Africa and Asia, the risk of human error during implementation, the risk of rapid global warming if an SRM project is abruptly stopped (the so-called termination shock risk), or the risk of military use of some technologies (the so-called dual use risk) (Robock 2016). Some SRM deployment scenarios can be quite anxiogenic, such as "parochial geoengineering", where current generations use SRM to secure short-term benefits for themselves by passing on much more serious long-term risks to future generations, or "predatory geoengineering", where one country chooses a particular form of geoengineering to disadvantage its geopolitical or economic rivals (Gardiner 2013; Gardiner and McKinnon 2024). Some scenarios have already been portrayed in Anthropocene fictions, especially in *Snowpiercer*, the graphic novel adapted into a film and then a television series (Lob et al. 2014). This climate fiction depicts a snowball Earth scenario, a post-apocalyptic future in which the planet is covered in ice because of the large-scale deployment of "CW-7", an artificial cooling substance that was meant to counter global warming but over-delivered. The possibility of unexpected consequences arising from this form of intentional climate change technique being added to the already unexpected consequences of the already existing climate change is one of the main uncertainties arising out of deploying such a measure: "the unpredictable effects of SRM are layered on top of the unpredictable effects of anthropogenic warming" (Preston 2012, 196).

The upshot is that the very measures taken to address environmental issues can contribute to eco-anxiety, as illustrated by poorly designed and/or implemented energy transition policies and SRM methods. Note that the relation between anxiety and technology is not unidirectional. Rapid technological development with uncertain consequences can be a driver of anxiety, but anxiety can also be a driver of technological innovation (Kalmbach 2025).[3]

This characterization of the object of eco-anxiety helps to provide some additional information on a subject briefly broached in Chapter 1: the newness of this ecological emotion. It is true that, throughout history, people have always been exposed to all sorts of risks and dangers from natural disasters. However, that does not make eco-anxiety a new label for an old reality. I see three main reasons for this, based on the account provided above.

First, the object of eco-anxiety is much more comprehensive than natural disasters that might have been observed in the past. Holocene problems were mainly characterized by pollution and environmental degradation that were spatially and temporally scaled to the size and dynamics of biotic communities and ecosystems. In contrast, Anthropocene problems represent global and intergenerational disruptions in the Earth system, such as climate change and biodiversity loss. This is a new kind of environmental degradation, which is best understood as a disruption in planetary-level systems, pushing the whole Earth system into a new state that is much less hospitable to humans and other species and threatens the very conditions of a flourishing human and non-human life.

The second reason why eco-anxiety is not simply a new way to name an old phenomenon is that its object also covers risks raised by the very technological measures that are envisaged to address these Anthropocene ecological problems, such as SRM. This global and intentional manipulation of the climate system is also a new form of technological intervention. It is a form of artificialization of the climate system that aims at changing some of the Earth's basic conditions to restore more stable climatic conditions for human societies (Bourban and Rochel 2021). Engineering our climate could turn the Earth into a giant artefact that would embody human intentional structure. This process of artificing the climate system comes with deep uncertainties that can be very anxiogenic.

The third reason is that many of the risks and dangers to which humans used to be exposed were imminent, which makes them an object of fear rather than anxiety. As we will see in the next subsection, fear is

---

[3] This has been observed for instance with the development of identification technologies in reaction to immigration or the development of surveillance technology in response to terrorism, but it might also be observed in the case of eco-anxiety, which can push for technological developments, for instance, in the domain of renewable technologies.

rather a defensive response triggered by clear and immediate dangers in one's surroundings that engage situation-appropriate behaviours to address the direct danger at hand, such as fight or flight behaviours. In contrast, anxiety is an emotional response triggered by threats that are real but whose likelihood and/or severity remain uncertain and that prompt epistemological behaviours of risk-assessment and risk-minimization. Our knowledge of future ecological risks is nowadays much more developed, thanks both to progress in environmental sciences and the digitalization of information. As explained in Chapter 1, this is also a key characteristic of the Anthropocene.

Environmental Awareness and Insecurity

Eco-anxiety is an emotional response to an acute awareness of ecological risks which leads to a state of insecurity. There are also two main elements here: (a) acute environmental awareness, and (b) a feeling of insecurity.

Regarding (a), eco-anxiety is usually triggered, at least initially, by the realization of the scope and severity of our planetary predicament. It is a normal response to a rapidly deteriorating ecological situation. Eco-anxiety is not an exaggerated anticipation of highly unlikely ecological risks and dangers; it is first and foremost a lucid reaction to an accurate empirical description of global environmental changes, some of which are already occurring. This is stressed by many studies on eco-anxiety. For instance, Caroline Hickman et al. (2021, e863) explain that "Although painful and distressing, climate anxiety is rational and does not imply mental illness". Likewise, Ashlee Cunsolo et al. (2020, e261) stress that both eco-grief and eco-anxiety represent "reasonable and functional responses to climate-related losses". Pierre-Eric Sutter et al. (2025, 19) add that "eco-anxiety (even when not very intense) reflects 'relevant' concerns about the environmental crisis, in the sense that they are based on proven scientific studies and facts about the state of the world". Finally, Holli-Anne Passmore et al. (2023, 140) explain that "Eco-anxiety is both an emotional and a rational response to very real threats. Our emotions help us survive—they indicate which situations and relationships require attention; in this case, our relationship with nature".

In other words, eco-anxiety is a *fitting emotional response* to an uncertain and threatening situation. Anthropocene problems come with a myriad of risks and dangers, and it is fitting to be anxious when one realizes the range and severity of problematic uncertainty we are facing.

For instance, as discussed in Chapter 1, it is an unequivocal fact that human economic activities have warmed the atmosphere, ocean, and land, and that this has already led to widespread and severe impacts, such as heatwaves, heavy precipitation, droughts, and cyclones (IPCC 2021). The situation really is a cause for concern, and this is why environmental scientists, and especially climate scientists, are so eco-anxious. For instance, a recent survey found that many IPCC authors are suffering from eco-anxiety, with more than 60% of the respondents saying that they experience anxiety, grief, or other distress because of concerns over climate change. Eighty-two per cent report that they think they will see catastrophic impacts of climate change in their lifetime, and six in ten expect the world to warm by at least 3 °C above preindustrial levels by 2100 (Tollefson 2021). The survey on the state of eco-anxiety in France also links environmental awareness with eco-anxiety. The authors found that one of the key determining factors of eco-anxiety is an interest in the environment and environmental protection, with a strong correlation between the two: the more people express interest in the environment, the more eco-anxious they are—and vice versa (Sutter et al. 2025). This means that the level of interest in the environment is a good predictor of the level of anxiety a person is feeling.

The role of environmental awareness in eco-anxiety shows that certainty also plays an important role in this ecological emotion. There are many things that eco-anxious people do not doubt. They know that climate change is real, that it is caused by human activities, and that multiple current and future climate impacts are harmful for human and other sentient beings. They also know that the solutions implemented so far have been largely insufficient and inadequate to properly address the root causes of global environmental changes and that those solutions can in turn create new risks. They know that the situation is bad and will continue to get worse (even if it can potentially improve). They know that the world will become a more dangerous and hostile place to live in. They know that ecological risks exist, even if their likelihood and severity can be difficult to assess. The fact that the likelihood and severity of ecological risks are increasingly well known shows that the initial source of anxiety is not (problematic) uncertainty; it is rather the certainty that something bad will happen and, if nothing is done about it, it will become much worse. Otherwise, eco-anxiety would gradually decrease as the projections of future ecological impacts become more robust. The fact that these clearer projections are feeding eco-anxiety feelings, rather than reducing them, shows that *problematic certainty* is just as important as problematic uncertainty. This tension between certainty and uncertainty is constitutive of the definition and experience of eco-anxiety.

Regarding (b), eco-anxious people feel insecure. Eco-anxiety is a state of constant worry that young people, future generations, and non-human beings will live in a more dangerous world. Insecurity and worry are often linked with an emotion that is close to anxiety: that of *fear*, which is sometimes used interchangeably with anxiety. It makes sense to consider eco-anxiety as a form of fear oriented towards environmental risks. After all, as we saw above, the APA itself is using the notion of fear to define eco-anxiety—a "chronic fear of environmental doom" (Clayton et al. 2017, 29). Against the tendency to compartmentalize and strictly separate our emotional states, it is important to recognize that our emotional experiences in a time of climate change are often interwoven (Oele 2024). Our understanding of anxiety is composed of a tangle of concepts and experiences, such as uncertainty, risk, threats, dangers, and fear (Schapira et al. 2024).

That being said, it remains important to distinguish eco-anxiety from fear for two main reasons. First, while fear is a defensive response to a danger in one's direct environment, anxiety is a defensive response to a more diffuse risk to the global environment. Second, while fear leads to situation-appropriate responses that address the direct danger at hand, such as fight or flight behaviours, anxiety leads to more general risk minimization and epistemic behaviours more adapted to addressing uncertain threats, such as information gathering. Kurth (2018b, 33) sums up these two important differences in the following two definitions:

> *Core fear*. A defensive response that is triggered by clear and present dangers in one's surroundings, and that engage specific, situation-appropriate behaviors (e.g., fight/flight/freeze).
> *Core anxiety*. A defensive response that is triggered by threats and challenges that are unpredictable, uncontrollable, or otherwise uncertain in nature, and that prompts general patterns of risk-assessment and risk-minimization behavior.

One major reason why anxiety can also be categorized as a mood is that, in contrast with fear, its object is not necessarily clear and close in space and time—it can be distant in both. This is why it is so difficult to get rid of anxiety: if we do not know precisely what is making us anxious, it is difficult to deal with the threat (Freeman and Freeman 2012, 11–12). As Nicole Shea and Emmanuel Kattan (2020, 1) put it, "Fear of immediate danger has been replaced by anxiety over an uncertain future and shapeless though imminent catastrophes".

It may make sense to consider eco-anxiety a mood with no specific object or with a very vague object such as "environmental doom". According to this approach, eco-anxiety represents a general feeling of doom without a specific object of attention, or with an object that is not specifically ecological in nature, such as political or moral risks, which can be related to ecological risks but are not reducible to them. For instance, anxiety about the rise of authoritarianism in response to climate impacts could be part of the mood aspect of eco-anxiety. This generalized mood-based eco-anxiety is rooted in the anticipation of the end of the world as we know it, of a breakdown of order and loss of things we morally value, typically because of political instability or societal collapse. In such cases, the behavioural response to eco-anxiety might be different from the ones we will discuss in the next subsection and could, for instance, include the behaviours of "preppers" who prepare for a catastrophic disaster by stockpiling food, ammunition, and other supplies and focus on individual survival.

This interpretation of eco-anxiety as a mood state can help us to understand some forms and manifestations of eco-anxiety, but the approach I support here focuses on eco-anxiety as an emotional response to a specific kind of threat. Since the main object of eco-anxiety is ecological risks, there is indeed an inescapable vagueness in our experience of it. However, eco-anxious people do not necessarily perceive the future as shapeless, vague, or amorphous; the futures they have in mind can be all too clear, with specific extreme climate events (heatwaves, hurricanes, droughts, and so on), their consequences (human and animal suffering, economic shocks, pandemics, and so on), and potential effects of the measures to address them (the social and environmental impacts of renewable energy technologies and climate engineering technologies). For this reason, eco-anxiety can last longer than other emotions, as its object is nowhere near disappearing or being addressed. Quite the contrary: ecological risks are not only increasingly better understood and described in the scientific literature, they are also worsening and ever more widespread. The direct experience of impacts of environmental change is not a necessary condition for feeling eco-anxious; it can be sufficient to indirectly learn about ecological risks in scientific reports, books, documentaries, series, novels, or movies to become eco-anxious.

Behavioural Responses

Eco-anxiety can lead to different behavioural responses. The two main behaviours I will focus on here are (a) risk-assessment and risk-minimization behaviours and (b) discouragement and paralysis.

Regarding (a), eco-anxiety is not just an emotional response to the awareness of the scale and severity of ecological risks; it also represents an adaptive response to environmental problems and their impacts. This adaptive response can help with maintaining positive mental health, provided that it is transformed into a driving force for action, particularly in the service of the environmental transition (Sutter et al. 2025). Eco-anxiety can lead to environmental action that can be a source of psychological well-being, as long as the actions remain in line with one's perception and self-assessment of one's abilities and resources. More specifically, eco-anxiety can lead to behaviours to address ecological risks, such as risk assessment and risk minimization. Risk-assessment behaviours include, for instance, the gathering of information on the causes, the structure, and the effects of environmental risks. Risk-minimization behaviours include deliberation and reflection on how to address the causes and how to adapt to the impacts of environmental problems. These epistemic behaviours are concerned with good or accurate decision making, such as making effective lifestyle changes to reduce one's ecological footprint or adapt to climate impacts.

Charlie Kurth and Panu Pihkala (2022) explain this important point when they link eco-anxiety with *practical anxiety*. Practical anxiety is a form of anxiety focused on the appropriate thing to do, prompted by uncertainty about how one might influence potentially threatening events in the future (Kurth 2018b). Practical anxiety tends to translate into epistemic behaviours that aim to help us understand and respond to the threat in question. Conceived as practical anxiety about ecological matters, eco-anxiety leads to an engagement in reflection and deliberation about what to do about ecological threats and a motivation to act according to the findings of that deliberative process. It brings both sensitivity and responsiveness to uncertainty about how best to respond to environmental threats. Kurth and Pihkala (2022, 8) summarize this point in the following way: "eco-anxiety functions as an alarm—it sensitizes individuals to situations where they face a novel or difficult decisions [sic] about something of ecological value; and it prompts the cognitive engagement and motivation that can help them address the difficulty they face".

Given the value of practical anxiety in terms of sensitivity and responsiveness to problematic uncertainty, Kurth (2018b, 136) argues that "anxiety is an emotion we ought to cultivate". He qualifies this statement by explaining that "the point is that (practical) anxiety is something we should learn to feel at the right times and in the right ways", especially "in its more typical and moderate manifestations", for instance when anxiety is "well-regulated" (Kurth 2018b, 135–136).

Whether or not it makes sense to cultivate (practical) anxiety, it is important to refrain from doing the same with eco-anxiety. There are three main reasons for this. First, "typical" forms of eco-anxiety do not necessarily correspond to its "moderate manifestations". For instance, when we look at the ways children and young people experience eco-anxiety, most of them think that humanity is doomed, and they are exposed to chronic stressors that could have considerable and long-lasting negative effects on their mental health (Hickman et al. 2021; Hickman 2024). The same can be said about the people and communities who are the most vulnerable to environmental impacts and who experience a mix of eco-anxiety, eco-grief, eco-fear, and other ecological emotions.

Second, it remains unclear whether it is possible to effectively "regulate" eco-anxiety when it is experienced to a high degree. Eco-anxiety can develop into overwhelming and disempowering emotions such as terror, horror, panic, grief, and other feelings of loss, which can lead in turn to feelings of frustration, despair, and depression. As Hickman (2020, 416) stresses, "Understandably in the face of such powerful and conflicting feelings defences can be easily triggered leading people to then feel numb or dissociate". Eco-anxiety is a source of distress and suffering to many, with a profound impact on to day-to-day functioning, and for this reason, it should not be cultivated, encouraged, or inculcated.

Third, risk-assessment and risk-minimization strategies might both turn out to be counterproductive and increase eco-anxiety levels. In the case of information gathering to assess ecological risks, the outcome might be that the problems seem so overwhelming that the eco-anxious person feels even worse for knowing more about environmental problems and their potentially catastrophic impacts. Think for instance here about the planetary boundaries that have been crossed or the tipping points in the climate system discussed in Chapter 1 or the fact that we are currently on a Hothouse Earth pathway. In the case of risk-minimization behaviours, the outcome might be that individual actions in the face of massive problems such as climate change and biodiversity loss seem pointless. Think for instance here about the size of an individual's or even a community's carbon footprint in comparison with the carbon footprint of a carbon major. In short, epistemological responses to eco-anxiety may backfire and further strengthen feelings of anxiety.

This leads us to (b): eco-anxiety can lead to a form of paralysis. The avalanche of data on the deteriorating state of the world in scientific and newspaper articles and documentaries can lead to "infowhelm", which prompts passivity and inaction. As Sarah Jaquette Ray (2020, 35) stresses, "Doomsayers can be as much a problem for the climate movement as deniers, because they spark guilt, fear, apathy, nihilism and ultimately inertia". Eco-anxiety can lead to a generalized feeling of discouragement regarding the future and what we can do about it, both individually and collectively. As Marjolein Oele (2024, 8) puts it, eco-anxiety "can become so stifling as to isolate us and deprive us of any action or choice".

Global environmental changes can lead to two different forms of inaction. The first typically appears when self-regulatory mechanisms of moral disengagement are activated. These psychological mechanisms allow people to "rationalise their reprehensible behaviour, and thus [permit] weakness of will and/or self-interested desires to thwart their moral motivation to abide by their moral judgement" (Peeters et al. 2019). They include discrediting evidence of harm, advantageous comparison, diffusion of responsibility, displacement of responsibility, unreasonable doubt, selective attention, and delusion. I shall call this *inaction as psychological defence*, inaction as a means to avoid the discomfort that comes with questioning our beliefs, convictions, and lifestyles.

The second form of inaction usually emerges after one has become aware of the state of the planet because of a direct experience of ecological impact or because of an indirect source such as a book, a report, or a documentary. Here, the facts are not denied, misrepresented, or underestimated: the gravity and the emergency of the situation are acknowledged. However, this knowledge can lead to helplessness, powerlessness, and resignation. As the APA stresses in the above-mentioned report, the psychological responses to climate change, such as conflict avoidance, fatalism, fear, helplessness and resignation are growing. These responses are keeping us, and our nation, from properly addressing the core causes of and solutions for our changing climate, and from building and supporting psychological resiliency. (Clayton et al. 2017, 4)

In this case, we move from inaction as defence to *inaction as paralysis*: it is no longer a problem of self-protection, but of discouragement, which is associated with the feeling that the issues at hand are just too massive and out of control for us to be able to address them.

Importantly, both forms of inaction can help with understanding why environmental action at the individual and collective levels remains largely insufficient to address global environmental problems such as climate change. Part of the problem can be explained by environmental indifference, which is a real issue. Another part of it could well be due to the fact that many are well aware of the severity of environmental problems but still do not act on it, either because their eco-anxiety has pushed them to activate psychological defence or moral disengagement mechanisms or because it has led them to a situation where they feel overwhelmed and discouraged and have eventually become paralysed. This can typically happen in cases where eco-anxious people feel like they cannot have any influence on the course of global environmental changes. In other words, a growing feeling of eco-anxiety can perfectly well coexist with widespread inaction against climate change and other environmental problems.

A Medical Condition?

Eco-anxiety is associated with mental health outcomes such as psychological distress, depression, insomnia, and stress symptoms (Boluda-Verdú et al. 2022; Cosh et al. 2024). Individuals can experience eco-anxiety at subclinical levels, but they can also experience impairment in day-to-day life related to eating, concentrating, work, school, sleeping, spending time in nature, playing, and/or having fun and relationships. Eco-anxiety can lead to mental disorders. The most severe mental health effects of climate change include PTSD, depression, the exacerbation of psychotic symptoms, suicidal ideation, and even suicide completion (Cunsolo et al. 2020). Since traditional forms of anxiety disorders can be related to environmental factors, it is to be expected that eco-anxiety can lead to psychological disorders (Bourban 2023). Phobias, an excessive or unreasonable fear of a specific object or situation, include natural environment phobias: excessive and unreasonable fear of storms, water, and fire can be exacerbated by climate impacts such as hurricanes, sea level rise, and megafires. Likewise, one possible factor contributing to PTSD is natural disasters.[4] The APA warns about these severe impacts of climate change on mental health in the 2021 edition of the report cited above:

> Climate change-fueled disaster events impact individual mental health and include trauma and shock, PTSD, anxiety and depression that can lead to *suicidal ideation* and *risky behavior*, feelings of abandonment, and physical health impacts. (Clayton et al. 2021, 6 – emphasis original)

Even if eco-anxiety can lead to such deleterious mental health issues, eco-anxiety does not represent a medical condition in itself. One reason for this is that most forms of eco-anxiety appear to be non-clinical (Pihkala 2020). Another reason is that eco-anxiety is not listed as a medical condition in the *Diagnostic and Statistical Manual of Mental Disorders* (DSM). Actually, many mental health professionals explain that it is important that eco-anxiety remains excluded from this manual as it would lead to pathologizing an emotion that comes from an accurate understanding of the severity of the ecological issues we are currently facing (Wray 2022, 21). Considering eco-anxiety as a mental health problem would indeed imply that people who are having a reasonable reaction to the ubiquitous manifestations of global environmental changes are mentally ill. This is why many scholars stress that eco-anxiety is a normal and healthy emotional response and should not be medicalized (Bhullar et al. 2022; Clayton 2020; Dodds 2021).

However, there might also be unintentional consequences to not treating eco-anxiety as a serious mental health problem for two main reasons (Van Valkengoed 2023). First, if eco-anxiety is only a normal and healthy response to global environmental changes, there would be no reason to try to find ways to cope with eco-anxiety or to mitigate it. However, for many people experiencing this ecological emotion, eco-anxiety can be very burdensome, with significant impacts on their daily

---

[4] As Cosh et al. (2024) stress, more research is needed to establish a clear link between eco-anxiety and PTSD, phobias, and other disorders such as GAD.

lives. This is why it is important to keep looking for possible ways to alleviate eco-anxiety. Second, studies have shown that eco-anxiety can be a source of motivation to engage in environmental action, such as reducing one's carbon footprint and engaging in collective action. If eco-anxiety is considered as normal and healthy, then one implication could be that we should find ways to instigate eco-anxiety in people who are not yet eco-anxious and sustain or even increase eco-anxiety levels in people who are already eco-anxious to reinforce action against environmental changes. Again, this might have severe effects on people's mental health, which should not be traded off for environmental protection as it represents a human right (see the next section).

How can this tension between pathologizing an essentially normal human response to a rapidly deteriorating ecological situation and being able to offer appropriate care to people in distress be solved? This is a complex question, and I cannot fully answer it here, but a good starting point is to draw on clinical case studies and research to distinguish between different ranges of feeling eco-anxiety, which come with different manifestations and mitigation strategies (Table 2.1).

Table 2.1 brings to the forefront a crucial point that has remained mostly implicit so far: the distinction between different *degrees of eco-anxiety*. The idea that eco-anxiety evolves along a spectrum or continuum has been stressed in recent systematic reviews of the literature on the topic (e.g. Cosh et al. 2024, 15; Léger-Goodes et al. 2022, 3). In mild and medium forms, it can push us to address environmental issues both at the individual and collective levels. In significant and severe forms, it can be paralysing and cause serious symptoms, such as insomnia, depressive episodes, substance abuse, self-harming, and difficulty maintaining functioning (Wray 2022, 56). Eco-anxiety can also be linked to reproductive anxiety, a reluctance to have children that derives from environmental concerns (Boluda-Verdú et al. 2022; Dillarstone et al. 2023; Wray 2022, 7–8, 80), and eco-nihilism, the idea that we should erase ourselves because we are so bad for the planet (Ray 2020, 40).

Table 2.1, which is based on years of experiences, observations, and research in psychotherapy, also shows that eco-anxiety can be both motivating and demotivating, both a resource and an obstacle to environmental action. This is also reflected in the literature: while some studies find that eco-anxiety encourages environmental action (Heeren et al. 2022; Whitmarsh et al. 2022), other studies do not find any significant correlation (Clayton and Karazsia 2020) or that eco-anxiety inhibits

Table 2.1 Degrees of eco-anxiety (1). Adapted from Hickman (2020, 417–418) and Wray (2022, 57–59)

Range of feeling	Manifestations	Mitigation strategies
Mild	Tendency to feel upset and to believe that only other people have the solutions to ecological problems, but this feeling is not constant. Painful feelings such as depression or despair are usually avoided	Individual and local actions, typically in terms of diets and recycling. Also a tendency to trust other people to find solutions and to be reassured by them
Medium	Tendency to feel more frequently upset and to start doubting that others can find solutions to ecological problems, but still a belief in the capacity of experts to take care of the situation. Feelings of discomfort and even distress have more strength	Lifestyle changes such as reduction in flying and meat consumption, but these changes are usually minimal. Can be reassured by discussions with others
Significant	Tendency to feel upset and in distress on a daily basis, with minimal psychological defences against guilt, grief, and fear. Fears about climate change are linked with fears of societal collapse. Increase in feeling of guilt and shame in relation to children and grandchildren	Anxiety is much harder to mitigate by reassurance; there is little faith in the capacity of other people to find solutions. The best way to reduce anxiety to a manageable level is through group actions such as activism and campaigning. Lifestyle changes are also much more significant, with commitments to stop flying and even to not having children (because of reproductive anxiety)
Severe	Tendency to experience sleep disruption and to struggle to enjoy any aspect of life because of intrusive thoughts. No psychological defence against the feeling of anxiety and a strong belief that societal collapse will take place in the future. No trust in others to solve environmental problems and sometimes inability to manage emotional responses. At most extreme, thoughts of suicide or having to kill one's children to save them from a violent death	One of the only ways to find personal security is the sense of belonging to a group, typically by joining group activism

climate action (Stanley et al. 2021). A major reason for this is that the relation between eco-anxiety and environmental action is influenced by the degree of eco-anxiety one feels: while overwhelming levels of anxiety could paralyse people, moderate levels of anxiety may be conducive to environmental action.

The relation between eco-anxiety and environmental action is not necessarily direct: it can be *mediated*, typically through other ecological emotions or through cognitions (Van Valkengoed et al. 2023, 259). For instance, while anger could turn eco-anxiety into action (Stanley et al. 2021), shame could lead to perceiving the self as a failure, which could lead to withdrawal and the avoidance of situations where action is possible (Swee et al. 2021). Also, people's perception of efficacy and responsibility could influence their motivation to act or not to act (Steg and De Groot 2010). Chapter 3 argues that ecological citizenship can strengthen this connection between eco-anxiety and environmental action, focusing especially on mild and moderate levels of eco-anxiety.

Table 2.2 proposes a slightly more detailed analysis of the different levels of eco-anxiety. It shows that eco-anxiety is a continuum with progressive levels and manifestations. To get an understanding of what it feels like to be eco-anxious, it is important to qualify this ecological emotion with an adjective expressing its intensity. This is why eco-anxiety should be conceived as a "progressive form of psychological distress" (Sutter et al. 2025, 9) moving from positive mental health (psychological well-being) to forms of psychopathologies (or mental illnesses). If emotional regulation and an approach to environmental action adapted to the individual's capacities are found, it can become an adaptive process in the face of ecological risks and develop favourably. However, if eco-anxiety becomes chronic or intensified and nothing is done to regulate or mitigate it, it can develop unfavourably. This is well illustrated by the risk threshold and the danger threshold. The first threshold indicates that the person concerned presents a very high level of eco-anxiety and should consult a psychotherapist to alleviate and treat their psychological distress. The second threshold indicates that the person concerned is likely to move from a state of psychological distress to an associated psychopathology such as depression or anxiety disorders.

It is important to stress that different levels of eco-anxiety are not experienced in a static way. People have fluctuating levels of eco-anxiety, depending on multiple external factors that can influence their emotional response to ecological threats, such as exposure to news cycles and

Table 2.2   Degrees of eco-anxiety (2). Adapted from Sutter et al. (2025)

Intensity category	Manifestations
Not at all eco-anxious	No manifestation of eco-anxiety
Very slightly eco-anxious	Having almost no worries about environmental problems and being almost never stressed and only slightly concerned by negative environmental information
Slightly eco-anxious	Being receptive to environmental problems and increasingly preoccupied with environmental information; there is no threat to mental health, but signs of distress are starting to show
Moderately eco-anxious	Showing more intense or chronic symptoms of eco-anxiety as a result of being particularly preoccupied by environmental information; this can cause stress peaks and sleep deprivation but does not pose a lasting threat to mental health
Strongly eco-anxious	Showing much more intense and chronic symptoms of eco-anxiety practically every day; environmental information becomes much more stressful and frightening, which starts posing a threat to mental health
Very strongly eco-anxious (Risk threshold)	Showing very intense, chronic, and long-lasting eco-anxiety symptoms and being very frequently preoccupied by environmental information, to the point of being traumatized by it; not being able to regulate one's feelings of eco-anxiety, with the emergence of dark thoughts about the future and even suicidal ideas
At risk of depression and anxiety disorders (Danger threshold)	Being constantly worried about environmental problems and being no longer able to control this worry. Being extremely frightened by the prospects raised by environmental problems, especially with a permanent fear of death, particularly concerning the future of species threatened with extinction. Seeing no way out of environmental problems and regularly having dark thoughts, including suicidal ideas

changes in election cycles. It is totally possible to move rapidly across these different categories, to change category from day to day, and even to experience different degrees of eco-anxiety within the same day. It can therefore be difficult to locate oneself on this scale and to know if the risk or danger threshold has been crossed, but the different forms of manifestations of eco-anxiety can help one to get a general idea of which degree of eco-anxiety one is feeling at a given moment. This scale needs to be further developed, and mental health professionals could use an improved version of it to provide support to eco-anxious people.

An Ethical and Political Problem

There are three main reasons why eco-anxiety is an ethical and political problem: it can be a source of moral injury, it can be threat to or violation of a basic human right, and it represents an emotional injustice.

First, eco-anxiety can lead to *moral injury*. Moral injury corresponds to the psychological aftermath experienced when one perpetrates or witnesses actions that violate one's core moral beliefs, typically in a situation of betrayal of justice by a person of authority (Griffin et al. 2019). While moral injury usually occurs in the context of military-related issues, it has recently been discussed in the context of eco-anxiety, especially regarding the role of governments in addressing environmental problems and their impacts on people's mental health.

The largest study to date on eco-anxiety, which surveyed 10,000 participants in 10 countries, stresses that "climate distress in children and young people can be regarded as unjust and involving moral injury. Young people's awareness of climate change and the inaction of governments are seen here to be associated with negative psychological sequelae" (Hickman et al. 2021, e871). When discussing their findings, the authors of this empirical study use the normative concepts of injustice and moral injury to explain that children and young people are not only anxious about the state of the planet but also experiencing confusion, betrayal, and abandonment because of the inaction of government in the face of environmental problems, to the point that some are turning to legal actions against their governments because of their failure to protect young citizens and their future (Salas et al. 2019).

There are two main aspects at play in the moral injury caused by eco-anxiety. First, there is a failure on the part of governments—and more generally adults—to act on environmental problems to protect the future conditions of life of children and young people. Hickman et al. (2021, e864) frame this first aspect as "a failure of ethical responsibility to care". Second, there is a failure on the part of governments and adults to acknowledge the validity of the feelings children and young people have in the face of environmental changes. All too often, government officials and other policymakers have dismissed, ignored, disavowed, or simply neglected these feelings. Even when they have not, they have simply made speeches and promises, without acting upon them.

The moral injury caused by eco-anxiety gives rise to a political responsibility to remedy this special kind of injury. Policymakers should (1) seriously acknowledge the negative feelings of young people and recognize their validity, and (2) place environmental action at the top of the political agenda and at the centre of policymaking. Here, policymakers, and especially government officials, have a double responsibility: they are responsible for the effects of their political inaction on the mental health of children and young people, and they are responsible for addressing these effects by helping those who suffer from eco-anxiety (Bourban 2024). This is also stressed by Navjot Bhullar et al. (2022, e383), who do not use the terminology of moral injury but still convey a strongly normative message when they write the following:

> It is the moral imperative and ethical responsibility of today's adults, especially those in positions of influence, to ensure adequate support and resources are available for young people as needed and to acknowledge their anxiety as a valid emotional response to a very real threat to their future, without being told that something is wrong with them.

Policymakers can implement different measures to fulfil their responsibilities (I draw here on some ideas proposed by Sutter et al. 2025). First, they can implement policies to raise awareness of eco-anxiety and its effects on mental health. Eco-anxiety remains a little-understood phenomenon among the general public, especially regarding its potential consequences on people's psychological well-being. Awareness policies could rely on communication campaigns on what eco-anxiety is, how it manifests itself, and what can be done to deal with its negative side effects such as stress, worry, and paralysis. Importantly, these policies should avoid contributing to the stigmatization of eco-anxious people and stress the positive side of eco-anxiety, presenting it as a normal reaction and adaptive response to ecological threats. Second, policymakers should subsidize research into eco-anxiety to develop our understanding of the factors that generate high levels of eco-anxiety and to find the most effective actions to regulate eco-anxiety. Third, they should invest in training for mental health professionals on how to deal with eco-anxiety in order to prevent its most extreme forms, for instance, by developing a network of coordinators trained in eco-anxiety issues. Fourth, they should further develop energy transition policies and encourage their citizens to participate in this transition, typically by making environmental actions more available and affordable (for instance through higher subsidies for renewable energy or by making public transportation cheap or even free).

The second reason why eco-anxiety is an ethical and political issue is that it also represents a *threat to or violation of the human right to health*. An influential definition of this right is provided by the International Covenant on Economic, Social, and Cultural Rights, which affirms "the right of everyone to the enjoyment of the highest attainable standard of physical and mental health" (UN 1976 art. 12.1). The advantage of this definition is that it focuses on both physical and mental health; the disadvantage is that it represents a maximalist interpretation of the right to health, which might be unrealistic in many cases where the "highest attainable standard" is not reachable or is only reachable with a very high level of resources that then could not be allocated to other important human rights or moral values.[5] For this reason, and following Simon Caney's account of human rights (Caney 2009), I interpret this right minimally and negatively as the right of everyone not to have their physical and mental health threatened by the actions of individual and collective agents.

---

[5] This formulation of the right to health is ambiguous. The expression "highest attainable standard" could be taken to mean the standard that can be attained by anyone (which would allow for interpersonal comparisons, for instance, with people who are much healthier), or it could be understood as the standard attainable for a particular individual (i.e. everyone should have access to health care that allows them to be as healthy as they *personally* can be). Even if we opt for the most charitable interpretation, which avoids the problem of interpersonal comparisons, this definition would still remain problematic, as many individuals have no realistic chance of reaching even the highest attainable standard of physical and mental health that they personally *can* achieve, because of economic or institutional constraints. This does not mean that one should not strive to get closer to this standard; my point is rather that a more modest formulation of the right to health is more likely to be applicable in the non-ideal circumstances of the real world and therefore more likely to be endorsed by the individuals and institutions responsible for protecting it.

Now, in the climate change literature, there has been a tendency to focus on the threats of climate change on physical health, typically by discussing the increase in the number of people suffering from disease and injury from heatwaves, floods, storms, fires, and droughts, for instance, through the increase in the number of people at risk of malaria and dengue or through the increase in the frequency of cardiorespiratory diseases (Bell 2011, 2013; Bourban 2014; Caney 2009). There are good reasons for this: the impacts of climate change on physical health are among the most ethically worrying. However, it is crucial not to neglect the effects of climate change and other environmental problems on mental health. In its latest assessment report, the IPCC has started to do so. In addition to impacts on physical health such as extreme heat events resulting in human mortality and morbidity, the authors add that "some mental health challenges are associated with increasing temperatures (*high confidence*), trauma from weather and climate extreme events (*very high confidence*), and loss of livelihoods and culture (*high confidence*)" (IPCC 2022, 11 - emphasis original). These findings are complemented by the two reports by ecoAmerica and the APA, which state that "The health, economic, political, and environmental implications of climate change affect all of us. The tolls on our mental health are far reaching. They induce stress, depression, and anxiety; strain social and community relationships; and have been linked to increases in aggression, violence, and crime" and that "Concern about climate change coupled with worry about the future can lead to fear, anger, feelings of powerlessness, exhaustion, stress, and sadness, referred to as ecoanxiety and climate anxiety" (Clayton et al. 2017, 4; 2021, 6). Importantly, the reports stress that children and young people are among the most exposed to these harmful impacts on mental health.

This might not seem sufficient to see a violation of or even a threat to the right to health of eco-anxious people. This objection has some truth to it, as not all eco-anxious people have their right to health jeopardized. However, in strong and very strong forms of eco-anxiety, it becomes difficult to maintain this objection. Coming back to Tables 2.1 and 2.2, it is safe to say that in most cases of high levels of eco-anxiety, there is a high risk of violation of the human right to mental health. This is the case, for instance, when eco-anxious people feel upset and/or in distress every day and have minimal psychological defences to deal with their emotions and mitigate their anxiety; when they are sleep deprived and struggle to enjoy any aspect of life because of intrusive thoughts; when they strongly believe that societal collapse will take place in the future; and/or when they start having thoughts of suicide or even thoughts of killing their own children to save them from a violent death. This is also confirmed by the two reports by ecoAmerica and the APA:

> Although the psychological impacts of climate change may not be obvious, they are no less serious because they can lead to disorders, such as depression, antisocial behavior, and suicide. Therefore, these disorders must be considered impacts of climate change as are disease, hunger, and other physical health consequences (Clayton et al. 2017, 7)

> Climate change-fueled disaster events impact individual mental health and include trauma and shock, PTSD, anxiety and depression that can lead to *suicidal ideation* and *risky behavior*, feelings of abandonment, and physical health impacts. (Clayton et al. 2021, 6 - emphasis original)

If the human right to health is defined as the right of everyone not to have their physical and mental health threatened by the actions of individual and collective agents, then the right to health of people who are severely or very severely eco-anxious is seriously threatened by anthropogenic climate change. This is a serious ethical problem, as the right to health is usually considered a basic right, that is, a right that is a condition of possibility for all other human rights, such as the right to education or the right to political participation. As Henry Shue (1996, 19) explains, since their "enjoyment is essential to the enjoyment of all other rights", basic rights are "everyone's minimum reasonable demands upon the rest of humanity". Once again, children and young people fall into a category that is particularly at risk, as their exposure to the chronic stressors of climate change and government inaction could have considerable, long-lasting, and incrementally negative implications for their mental health (Hickman et al. 2021). This should push environmental studies scholars, and especially environmental justice and climate justice researchers, to take the impact of environmental change on mental health (more) seriously.

In addition to children and young people, a key category that is exposed to this risk of violation of the right to (mental health) is that of people directly exposed to stress and trauma from more frequent and severe extreme weather and climate events. This category covers especially those who rely closely on the land and land-based activities, such as Indigenous people and farmers, but also, more generally, low-income households (IPCC 2022). These people and communities are more at risk of having severe mental health issues related to climate change because of their lower capacity to adapt, as they often lack the financial or social capacity to cope, manage, or recover from environmental hazards and climate stress (Ingle and Mikulewicz 2020). Eco-anxiety is intensely experienced by those most exposed to climate impacts because of physical ecological losses and their catastrophic effects on traditional ways of life and cultures, disruption of environmental knowledge systems, and the resulting feelings of identity loss and anticipated future losses of place, land, species, and culture (Cunsolo and Ellis 2018; Cunsolo et al. 2020).

The third reason why eco-anxiety is an ethical and political issue is that it represents a problem of *emotional justice*. One major reason why climate change is unjust is that those who have contributed the least to causing it are the most vulnerable to its impacts on physical and mental health (Verlie 2024). When small-scale farmers, Indigenous people, children, and young people feel eco-anxious because of their experience or anticipation of ecological risks, they find themselves in a situation of emotional injustice. The idea here is that climate change is unjust not only because of an unfair allocation of climate burdens, in terms of mitigation, adaptation, and loss and damage policies; it is unjust because of an unfair allocation of emotional burdens, with those who feel the most anxious about climate change and its impact being the least responsible for GHG-intensive emitting activities. In other words, the climate burdens that need to be fairly allocated do not only cover mitigation, adaptation, and loss and damage burdens, but also emotional burdens. Such burdens are not restricted to eco-anxiety but also include other threat-related ecological emotions, such as Anthropocene horror, as well as sadness-related emotions, such as eco-grief.

Avoiding moral injury and avoiding the violation of the human right to (mental) health represent objectives of harm-avoidance justice; the most important goal here is twofold: (a) to minimize or avoid harm to people (for instance by aggressively mitigating global GHG emissions) and (b) repair or at least compensate the harms that have not been avoided in the past (for instance, by training mental health professionals to identify severe cases of eco-anxiety and treat them adequately). In contrast, addressing emotional injustice is a fairness-related problem: the objective is to allocate the emotional burdens of global ecological problems more fairly between those who are mainly responsible for creating, maintaining, and/or worsening environmental problems, who should feel more worried about the impacts of their lifestyles, policies, and business models, and those who are the most exposed to environmental impacts, who should be provided with more adequate, accessible, and affordable support by mental health professionals. In other words, the notion of emotional (in)justice adds the idea of fairness to the normative picture provided by the notions of moral injustice and human right violation.

To conclude this chapter, it is important to make two distinctions: between healthy and unhealthy forms of eco-anxiety and between the rational and the irrational sides of eco-anxiety. On the one hand, it is crucial not to pathologize an emotion that arises out of an adequate and healthy response to an acute awareness of ecological risks. On the other hand, to avoid underestimating the severity of the effects eco-anxiety can have on mental health or encouraging the development of this emotion in those who do not feel it or only feel it in mild or moderate forms, it is equally crucial to highlight the pathologies eco-anxiety can lead to, such as PTSD, phobias, depression, risky behaviour, suicidal ideation, and even suicide completion. These pathologies are, however, not part of eco-anxiety per se; they are possible effects of eco-anxiety when it crosses the danger threshold and leads to severe side effects. Eco-anxiety is not intrinsically pathological, but it can lead to mental health disorders. This is why it should be neither underestimated nor cultivated.

Turning to the second distinction, eco-anxiety arises out of a rational response to the threats related to global environmental changes. It is a fitting emotional reaction to ecological risks. This does not mean that eco-anxiety is exclusively rational; there could also be an irrational side to eco-anxiety, for instance if it is based on an exaggeration of the probabilities and/or the severity of ecological risks. This means that someone can become eco-anxious through assessing the risks wrongly—for instance, believing that the world will warm by over 4 °C in the next couple of decades—and be terrified for this reason. The emotional response of this person would be based on an irrational assessment of the risks, but the emotion would still correspond to eco-anxiety—although it would also draw closer to other threat-related ecological emotions, such as Anthropocene horror and global dread. This has significant implications for people's disposition to remain hopeful in the face of Anthropocene problems, a topic that will be discussed in detail in the next chapter, to which I am turning now.

REFERENCES

Albrecht, Glenn. 2011. 'Chronic Environmental Change: Emerging "Psychoterratic" Syndromes'. In *Climate Change and Human Well-Being*, edited by Inka Weissbecker. International and Cultural Psychology. New York: Springer. https://doi.org/10.1007/978-1-4419-9742-5_3.

Albrecht, Glenn. 2019. *Earth Emotions: New Words for a New World*. Cornell University Press.

Andreucci, Diego, Gustavo García López, Isabella M. Radhuber, et al. 2023. 'The Coloniality of Green Extractivism: Unearthing Decarbonisation by Dispossession through the Case of Nickel'. *Political Geography* 107 (November): 102997. https://doi.org/10.1016/j.polgeo.2023.102997.

APA. 2000. *Diagnostic and Statistical Manual of Mental Disorders: DSM-IV-TR*. American Psychiatric Press.

Bell, Derek. 2011. 'Does Anthropogenic Climate Change Violate Human Rights?' *Critical Review of International Social and Political Philosophy* 14 (2): 99–124. https://doi.org/10.1080/13698230.2011.529703.

Bell, Derek. 2013. 'Climate Change and Human Rights'. *Wires Climate Change* 4 (3): 159–170. https://doi.org/10.1002/wcc.218.

Benatar, David. 2017. *The Human Predicament: A Candid Guide to Life's Biggest Questions*. Oxford: Oxford University Press.

Bergo, Bettina G. 2021. *Anxiety: A Philosophical History*. Oxford University Press.

Bhullar, Navjot, Melissa Davis, Roselyn Kumar, Patrick Nunn, and Debra Rickwood. 2022. 'Climate Anxiety Does Not Need a Diagnosis of a Mental Health Disorder'. *The Lancet Planetary Health* 6 (5): e383. https://doi.org/10.1016/S2542-5196(22)00072-9.

Boluda-Verdú, Inmaculada, Marina Senent-Valero, Mariola Casas-Escolano, Alicia Matijasevich, and María Pastor-Valero. 2022. 'Fear for the Future: Eco-Anxiety and Health Implications, a Systematic Review'. *Journal of Environmental Psychology* 84 (December): 101904. https://doi.org/10.1016/j.jenvp.2022.101904.

Bourban, Michel. 2014. 'Climate Change, Human Rights and the Problem of Motivation'. *De Ethica* 1 (1): 37–52. https://doi.org/10.3384/de-ethica.2001-8819.141137.

Bourban, Michel. 2023. 'Eco-Anxiety and the Responses of Ecological Citizenship and Mindfulness'. In *The Palgrave Handbook of Environmental Politics and Theory*, edited by Joel Jay Kassiola and Timothy W. Luke. Environmental Politics and Theory. Springer International Publishing. https://doi.org/10.1007/978-3-031-14346-5_4.

Bourban, Michel. 2024. 'Eco-Anxiety: A Philosophical Approach'. In *Anxiety Culture: The New Global State of Human Affairs*, edited by John Allegrante, Ulrich Hoinkes, Michael Schapira, and Karen Struve. Johns Hopkins University Press.

Bourban, Michel, and Johan Rochel. 2021. 'Synergies in Innovation: Lessons Learnt from Innovation Ethics for Responsible Innovation'. *Philosophy & Technology* 34 (2): 373–394. https://doi.org/10.1007/s13347-020-00392-w.

Bourg, Dominique. 2013. 'Dommages Transcendantaux'. In *Du Risque à La Menace : Penser La Catastrophe*, edited by Dominique Bourg, Pierre-Benoît Joly, and Alain Kaufmann. PUF.

Brophy, Hailie, Joanne Olson, and Pauline Paul. 2023. 'Eco-anxiety in Youth: An Integrative Literature Review'. *International Journal of Mental Health Nursing* 32 (3): 633–661. https://doi.org/10.1111/inm.13099.

Caney, Simon. 2009. 'Climate Change, Human Rights and Moral Thresholds'. In *Human Rights and Climate Change*, edited by Stephen Humphreys. Cambridge University Press.

Ceballos, Gerardo, Paul R. Ehrlich, and Peter H. Raven. 2020. 'Vertebrates on the Brink as Indicators of Biological Annihilation and the Sixth Mass Extinction'. *Proceedings of the National Academy of Sciences* 117 (24): 13596–13602. https://doi.org/10.1073/pnas.1922686117.

Clayton, Susan. 2020. 'Climate Anxiety: Psychological Responses to Climate Change'. *Journal of Anxiety Disorders* 74 (August): 102263. https://doi.org/10.1016/j.janxdis.2020.102263.

Clayton, Susan, and Bryan T. Karazsia. 2020. 'Development and Validation of a Measure of Climate Change Anxiety'. *Journal of Environmental Psychology* 69 (June): 101434. https://doi.org/10.1016/j.jenvp.2020.101434.

Clayton, Susan, Christie Manning, Kirra Krygsman, and Meighen Speiser. 2017. *Mental Health and Our Changing Climate: Impacts, Implications, and Guidance*. American Psychological Association & ecoAmerica.

Clayton, Susan, Christie Manning, Meighen Speiser, and Alison N. Hill. 2021. *Mental Health and Our Changing Climate: Impacts, Inequities, Responses*. American Psychological Association & ecoAmerica.

Coffey, Yumiko, Navjot Bhullar, Joanne Durkin, Md Shahidul Islam, and Kim Usher. 2021. 'Understanding Eco-Anxiety: A Systematic Scoping Review of Current Literature and Identified Knowledge Gaps'. *The Journal of Climate Change and Health* 3 (August): 100047. https://doi.org/10.1016/j.joclim.2021.100047

Cosh, Suzanne M., Rosie Ryan, Kaii Fallander, et al. 2024. 'The Relationship between Climate Change and Mental Health: A Systematic Review of the Association between Eco-Anxiety, Psychological Distress, and Symptoms of Major Affective Disorders'. *BMC Psychiatry* 24 (1): 833. https://doi.org/10.1186/s12888-024-06274-1.

Cunsolo, Ashlee, and Neville R. Ellis. 2018. 'Ecological Grief as a Mental Health Response to Climate Change-Related Loss'. *Nature Climate Change* 8 (4): 275–281. https://doi.org/10.1038/s41558-018-0092-2.

Cunsolo, Ashlee, Sherilee L. Harper, Kelton Minor, Katie Hayes, Kimberly G. Williams, and Courtney Howard. 2020. 'Ecological Grief and Anxiety: The Start of a Healthy Response to Climate Change?' *The Lancet Planetary Health* 4 (7): e261–e263. https://doi.org/10.1016/S2542-5196(20)30144-3.

Dillarstone, Hope, Laura J. Brown, and Elaine C. Flores. 2023. 'Climate Change, Mental Health, and Reproductive Decision-Making: A Systematic Review'. *PLOS Climate* 2 (11): e0000236. https://doi.org/10.1371/journal.pclm.0000236.

Dodds, Joseph. 2021. 'The Psychology of Climate Anxiety'. *Bjpsych Bulletin* 45 (4): 222–226. https://doi.org/10.1192/bjb.2021.18.

Freeman, Daniel, and Jason Freeman. 2012. *Anxiety: A Very Short Introduction.* Oxford University Press.

Gago, Tomás, Rebecca J. Sargisson, and Taciano L. Milfont. 2024. 'A Meta-Analysis on the Relationship between Climate Anxiety and Wellbeing'. *Journal of Environmental Psychology* 94 (March): 102230. https://doi.org/10.1016/j.jenvp.2024.102230.

Gardiner, Stephen M. 2013. 'The Desperation Argument for Geoengineering'. *Political Science and Politics* 46 (1): 28–33. https://doi.org/10.1017/S1049096512001424.

Gardiner, Stephen M., and Catriona McKinnon. 2024. 'Generationally Parochial Geoengineering: Early Warning Signs of a Basic Threat'. *The Institute for Futures Studies. Working Paper 2024:14*, 103–137.

Gasparatos, Alexandros, Christopher N.H.. Doll, Miguel Esteban, Abubakari Ahmed, and Tabitha A. Olang. 2017. 'Renewable Energy and Biodiversity: Implications for Transitioning to a Green Economy'. *Renewable and Sustainable Energy Reviews* 70 (April): 161–184. https://doi.org/10.1016/j.rser.2016.08.030.

Gregersen, Thea, Rouven Doran, Charles A. Ogunbode, and Gisela Böhm. 2024. 'How the Public Understands and Reacts to the Term "Climate Anxiety." *Journal of Environmental Psychology* 96 (June): 102340. https://doi.org/10.1016/j.jenvp.2024.102340.

Griffin, Brandon J., Natalie Purcell, Kristine Burkman, et al. 2019. 'Moral Injury: An Integrative Review'. *Journal of Traumatic Stress* 32 (3): 350–362. https://doi.org/10.1002/jts.22362.

Heeren, Alexandre, Camille Mouguiama-Daouda, and Alba Contreras. 2022. 'On Climate Anxiety and the Threat It May Pose to Daily Life Functioning and Adaptation: A Study among European and African French-Speaking Participants'. *Climatic Change* 173 (1–2): 15. https://doi.org/10.1007/s10584-022-03402-2.

Hickman, Caroline. 2020. 'We Need to (Find a Way to) Talk about ... Eco-Anxiety'. *Journal of Social Work Practice* 34 (4): 411–424. https://doi.org/10.1080/02650533.2020.1844166.

Hickman, Caroline. 2024. 'Eco-Anxiety in Children and Young People—A Rational Response, Irreconcilable Despair, or Both?' *The Psychoanalytic Study of the Child* 77 (1): 356–368. https://doi.org/10.1080/00797308.2023.2287381.

Hickman, Caroline, Elizabeth Marks, Panu Pihkala, et al. 2021. 'Climate Anxiety in Children and Young People and Their Beliefs about Government Responses to Climate Change: A Global Survey'. *The Lancet Planetary Health* 5 (12): e863–e873. https://doi.org/10.1016/S2542-5196(21)00278-3.

Hopster, Jeroen. 2021. 'What Are Socially Disruptive Technologies?' *Technology in Society* 67 (November): 101750. https://doi.org/10.1016/j.techsoc.2021.101750.

Ingle, Harriet E., and Michael Mikulewicz. 2020. 'Mental Health and Climate Change: Tackling Invisible Injustice'. *The Lancet Planetary Health* 4 (4): e128–e130. https://doi.org/10.1016/S2542-5196(20)30081-4.

IPCC. 2021. 'Summary for Policymakers'. In *Climate Change 2021: The Physical Science Basis. Contribution of Working Group I to the Sixth Assessment Report of the Intergovernmental Panel on Climate Change*, edited by V. Masson-Delmotte, P. Zhai, A. Pirani, et al. Cambridge University Press.

IPCC. 2022. 'Summary for Policymakers'. In *Climate Change 2022: Impacts, Adaptation, and Vulnerability. Contribution of Working Group II to the Sixth Assessment Report of the Intergovernmental Panel on Climate Change*, edited by H.-O. Pörtner, D. C. Roberts, E. S. Poloczanska, et al. Cambridge University Press.

Kalmbach, Karena. 2025. 'Fear and Technology in Modern Europe: A Call for Researching Fears as Drivers of Technology Development'. In *Anxiety Culture: The New Global State of Human Affairs*, edited by John Allegrante, Ulrich Hoinkes, Michael Schapira, and Karen Struve. Johns Hopkins University Press.

Kurth, Charlie. 2018a. 'Anxiety: A Case Study on the Value of Negative Emotion'. In *Shadows of the Soul: Philosophical Perspectives on Negative Emotions*, edited by Christine Tappolet, Fabrice Teroni, and Anita Konzelmann Ziv. Routledge, Taylor & Francis Group. https://doi.org/10.4324/9781315537467.

Kurth, Charlie. 2018b. *The Anxious Mind: An Investigation into the Varieties and Virtues of Anxiety*. MIT Press.

Kurth, Charlie, and Panu Pihkala. 2022. 'Eco-Anxiety: What It Is and Why It Matters'. *Frontiers in Psychology* 13 (September): 981814. https://doi.org/10.3389/fpsyg.2022.981814.

Lafontaine, Simon. 2022. 'Anxiety and the Re-Figuration of Action: Living in a Crisis-Shaped Present'. In *Eco-Anxiety and Planetary Hope*, edited by Douglas A. Vakoch and Sam Mickey. Springer International Publishing. https://doi. org/10.1007/978-3-031-08431-7_4.

Léger-Goodes, Terra, Catherine Malboeuf-Hurtubise, Trinity Mastine, Mélissa. Généreux, Pier-Olivier. Paradis, and Chantal Camden. 2022. 'Eco-Anxiety in Children: A Scoping Review of the Mental Health Impacts of the Awareness of Climate Change'. *Frontiers in Psychology* 13 (July): 872544. https://doi. org/10.3389/fpsyg.2022.872544.

Levenda, A.M., I. Behrsin, and F. Disano. 2021. 'Renewable Energy for Whom? A Global Systematic Review of the Environmental Justice Implications of Renewable Energy Technologies'. *Energy Research and Social Science* 71. Scopus. https://doi.org/10.1016/j.erss.2020.101837.

Lob, Jacques, Jean-Marc Rochette, and Benjamin Legrand. 2014. *Transperceneige: intégrale*. Casterman.

Miceli, Maria, and Cristiano Castelfranchi. 2005. 'Anxiety as an "Epistemic" Emotion: An Uncertainty Theory of Anxiety'. *Anxiety, Stress & Coping* 18 (4): 291–319. https://doi.org/10.1080/10615800500209324.

Mickey, Sam. 2022. 'Atmospheres of Anxiety: Doing Nothing in an Ecological Emergency'. In *Eco-Anxiety and Planetary Hope*, edited by Douglas A. Vakoch and Sam Mickey. Springer International Publishing. https://doi.org/ 10.1007/978-3-031-08431-7_3.

Murphy, Laura, and Nyrola Elimä. 2021. *In Broad Daylight: Uyghur Forced Labour and Global Solar Supply Chains*. Sheffield Hallam University Helena Kennedy Centre for International Justice. https://www.shu.ac.uk/-/media/ home/research/helena-kennedy-centre/projects/pdfs/evidence-base/in-broad-daylight.pdf.

Oele, Marjolein. 2024. 'Anxiety, Grief, and Trust in Times of Climate Change: A Phenomenology of Affective Constellations and Future Transformations in and beyond the Anthropocene'. *Comparative and Continental Philosophy*, June 3, 1–20. https://doi.org/10.1080/17570638.2024.2361409.

Ojala, Maria, Ashlee Cunsolo, Charles A. Ogunbode, and Jacqueline Middleton. 2021. 'Anxiety, Worry, and Grief in a Time of Environmental and Climate Crisis: A Narrative Review'. *Annual Review of Environment and Resources* 46 (1): 35–58. https://doi.org/10.1146/annurev-environ-012220-022716.

Oksala, Johanna. 2023. 'The Existential Threat of Climate Change: From Climate Anxiety to Post-Nihilist Politics'. *Environmental Philosophy* 20 (2): 191–214. https://doi.org/10.5840/envirophil2023919133.

Passmore, Holli-Anne., Paul K. Lutz, and Andrew J. Howell. 2023. 'Eco-Anxiety: A Cascade of Fundamental Existential Anxieties'. *Journal of Constructivist Psychology* 36 (2): 138–153. https://doi.org/10.1080/10720537. 2022.2068706.

Peeters, Wouter, Lisa Diependaele, and Sigrid Sterckx. 2019. 'Moral Disengagement and the Motivational Gap in Climate Change'. *Ethical Theory and Moral Practice* 22 (2): 425–447. https://doi.org/10.1007/s10677-019-09995-5.

Pihkala, Panu. 2020. 'Anxiety and the Ecological Crisis: An Analysis of Eco-Anxiety and Climate Anxiety'. *Sustainability* 12 (19): 7836. https://doi.org/10.3390/su12197836.

Preston, Christopher J. 2012. 'Beyond the End of Nature: SRM and Two Tales of Artificity for the Anthropocene'. *Ethics, Policy & Environment: A Journal of Philosophy and Geography* 15 (2): 188–201. https://doi.org/10.1080/215 50085.2012.685571.

Ray, Sarah Jaquette. 2020. *A Field Guide to Climate Anxiety: How to Keep Your Cool on a Warming Planet*. University of California Press.

Robock, Alan. 2016. 'Albedo Enhancement by Stratospheric Sulfur Injections: More Research Needed'. *Earth's Future* 4 (12): 644–648. https://doi.org/10.1002/2016EF000407.

Royal Society. 2009. *Geoengineering the Climate: Science, Governance and Uncertainty*. Royal Society. https://royalsociety.org/news-resources/publications/2009/geoengineering-climate/.

Salas, Renee N., Wendy Jacobs, and Frederica Perera. 2019. 'The Case of *Juliana v. U.S.*—Children and the Health Burdens of Climate Change'. *New England Journal of Medicine* 380 (22): 2085–2087. https://doi.org/10.1056/NEJ Mp1905504.

Schapira, Michael, Karen Struve, Ulrich Hoinkes, and John Allegrante. 2024. 'Introduction: Anxiety as a New Global Narrative'. In *Anxiety Culture: The New Global State of Human Affairs*, edited by John Allegrante, Ulrich Hoinkes, Michael Schapira, and Karen Struve. Johns Hopkins University Press.

Shea, Nicole, and Emmanuel Kattan. 2020. 'Anxiety Culture'. *EuropeNow: A Journal of Research & Art*.

Shue, Henry. 1996. *Basic Rights: Subsistence, Affluence, and U.S. Foreign Policy*. 2nd edn. Princeton University Press.

Shue, Henry. 2010. 'Deadly Delays, Saving Opportunities: Creating a More Dangerous World?' In *Climate Ethics: Essential Readings*, edited by Stephen Gardiner, Simon Caney, Dale Jamieson, and Henry Shue. Oxford University Press.

Sonter, Laura J., Marie C. Dade, James E. M. Watson, and Rick K. Valenta. 2020. 'Renewable Energy Production Will Exacerbate Mining Threats to Biodiversity'. *Nature Communications* 11 (1): 4174. https://doi.org/10.1038/s41 467-020-17928-5.

Sovacool, Benjamin K., Mari Martiskainen, Andrew Hook, and Lucy Baker. 2019. 'Decarbonization and Its Discontents: A Critical Energy Justice

Perspective on Four Low-Carbon Transitions'. *Climatic Change* 155 (4): 581–619. https://doi.org/10.1007/s10584-019-02521-7.

Stanley, Samantha K., Teaghan L. Hogg, Zoe Leviston, and Iain Walker. 2021. 'From Anger to Action: Differential Impacts of Eco-Anxiety, Eco-Depression, and Eco-Anger on Climate Action and Wellbeing'. *The Journal of Climate Change and Health* 1 (March): 100003. https://doi.org/10.1016/j.joclim.2021.100003.

Steffen, Will, Johan Rockström, Katherine Richardson, et al. 2018. 'Trajectories of the Earth System in the Anthropocene'. *Proceedings of the National Academy of Sciences* 115 (33): 8252–8259. https://doi.org/10.1073/pnas.1810141115.

Steg, Linda., and Judith. De Groot. 2010. 'Explaining Prosocial Intentions: Testing Causal Relationships in the Norm Activation Model'. *British Journal of Social Psychology* 49 (4): 725–43. https://doi.org/10.1348/014466609X477745.

Sutter, Pierre-Eric, Jean-Luc Bernaud, and Léonie Messmer. 2025. *Éco-Anxiété En France (Étude 2025)*. ADEME. https://librairie.ademe.fr/societe-et-politiques-publiques/8137-eco-anxiete-en-france.html.

Swee, Michaela B., Chloe C. Hudson, and Richard G. Heimberg. 2021. 'Examining the Relationship between Shame and Social Anxiety Disorder: A Systematic Review'. *Clinical Psychology Review* 90 (December): 102088. https://doi.org/10.1016/j.cpr.2021.102088.

Tollefson, Jeff. 2021. 'Top Climate Scientists Are Sceptical That Nations Will Rein in Global Warming'. *Nature* 599 (4 November): 22–24. https://doi.org/10.1038/d41586-021-02990-w.

UN. 1976. *International Covenant on Economic, Social and Cultural Rights*. General Assembly resolution 2200A (XXI).

Valkengoed, Van, and M. Anne. 2023. 'Climate Anxiety Is Not a Mental Health Problem. But We Should Still Treat It as One'. *Bulletin of the Atomic Scientists* 79 (6): 385–387. https://doi.org/10.1080/00963402.2023.2266942.

Valkengoed, Van, M. Anne, Linda Steg, and Peter De Jonge. 2023. 'Climate Anxiety: A Research Agenda Inspired by Emotion Research'. *Emotion Review* 15 (4): 258–262. https://doi.org/10.1177/17540739231193752.

Vaškovic, Petr. 2023. 'Philosophical Perspectives on Climate Anxiety'. In *Handbook of the Philosophy of Climate Change*, edited by Gianfranco Pellegrino and Marcello Di Paola. Handbooks in Philosophy. Springer International Publishing. https://doi.org/10.1007/978-3-031-07002-0_144.

Vaškovic, Petr, and Gabriela Vičanová. 2024. 'Anxiety, Hope and Meaning in Times of Ecological Crisis: An Existential-Phenomenological Perspective on Environmental Emotions'. *Human Studies* 47 (4): 771–791. https://doi.org/10.1007/s10746-024-09728-3.

Vazard, Juliette, and Charlie Kurth. 2022. 'Apprehending Anxiety: An Introduction to the Topical Collection on Worry and Wellbeing'. *Synthese* 200 (4): 327, s11229-022-03794-99. https://doi.org/10.1007/s11229-022-03794-9.

Verlie, Blanche. 2024. 'A Pedagogy for Emotional Climate Justice'. In *The Existential Toolkit for Climate Justice Educators: How to Teach in a Burning World*, 1st edn, edited by Jennifer Atkinson and Sarah Jaquette Ray. University of California Press. https://doi.org/10.2307/jj.14284466.

Whitmarsh, Lorraine, Lois Player, Angelica Jiongco, et al. 2022. 'Climate Anxiety: What Predicts It and How Is It Related to Climate Action?' *Journal of Environmental Psychology* 83 (October): 101866. https://doi.org/10.1016/j.jenvp.2022.101866.

Wray, Britt. 2022. *Generation Dread: Finding Purpose in an Age of Climate Crisis*. Alfred A. Knopf Canada.

CHAPTER 3

Ecological Citizenship: Addressing and Harnessing Eco-Anxiety

Abstract This chapter explains how to cope with eco-anxiety, both by mitigating its undesirable effects such as worry and stress and by harnessing its beneficial consequences such as risk-assessment and risk-minimization behaviours. It shows that ecological citizenship can serve as a normative framework that is relevant to navigating eco-anxiety as it provides coping strategies at the individual and collective levels. It investigates three of the main values of the ecological citizen. The first is carbon sobriety, a voluntary form of moderation applied to one's individual carbon footprint. The second is hope, a disposition that encourages a realistic outlook on an uncertain future and that helps with avoiding the pitfalls of optimism and pessimism. The third is courage, an ability to act to protect environmental value despite the risks involved. All three values, especially when taken together, can help eco-anxious people to face our planetary predicament without giving in to eco-despair.

Keywords Eco-Anxiety · Ecological citizenship · Environmental action · Good Hope · Radical Hope · Despair · Courage · Scenarios

Now that the contours of the notion of eco-anxiety are clarified and the different degrees of eco-anxiety have been presented, it is time to ask how to cope with eco-anxiety. The goal here is not to try to get rid of this ecological emotion: this is neither possible, nor, as we will see,

© The Author(s) 2026
M. Bourban, *Eco-Anxiety and Ecological Citizenship*,
https://doi.org/10.1007/978-3-032-03219-5_3

desirable. The objective is rather to learn to live with eco-anxiety, first by ensuring that it does not reach a degree where it becomes unmanageable and leads to pathological consequences, such as PTSD, phobia, and depression, second by mitigating its undesirable side effects, such as worry, stress, and paralysis, and third by harnessing its beneficial effects, such as information gathering and risk-assessment behaviours as well as reflection and deliberation, to respond to environmental threats.

Possible means to address eco-anxiety include psychological therapy, including cognitive-behavioural therapy (CBT), and medication through anxiolytics and antidepressants. CBT may be especially well suited, as its effectiveness as a treatment for anxiety disorders has been proven, even though there is still room for improvement (Hoffman and Smits 2008). These measures, which rely on the professional assistance of psychologists and psychiatrists, are especially appropriate in cases of severe and very severe eco-anxiety, but they can also be useful in coping with some effects of lower degrees of eco-anxiety, which can also affect individuals' well-being. Getting psychological help from a mental health professional is crucial, and this is why it is so important for such professionals to develop their skills and knowledge relating to environmental factors, thereby including consideration of eco-anxiety in clinical practice (Cosh et al. 2024). Lifestyle changes, such as increased physical exercise, healthy diets, and yoga can also help (Manzoni et al. 2008), but they will never represent a substitute for professional support.

Although psychological therapy and medication can contribute to reducing eco-anxiety levels when they pass the risk threshold and the danger threshold and therefore pose a significant threat to mental health (see Chapter 2), the psychological and psychiatric approach is not a silver bullet. There are three main reasons for this. First, access to mental health services and resources is insufficient. The limited availability and affordability of appropriate mental health care have been exacerbated by the COVID-19 pandemic (Lattie et al. 2022). In a context of scarcity of adequate psychological and psychiatric services and resources, it is important to look for alternative approaches. Second, there is a risk of medicalizing eco-anxiety if we approach it exclusively or mainly from a psychiatric angle. This is something that has been observed in the case of anxiety: while the efficacy of psychotherapy such as CBT is well documented in the literature (e.g. APA 2013), medication remains the most common treatment for anxiety.[1] As Mariana Alessandri (2024, 140) stresses, since pharmaceuticals started flooding the market in the 1950s with the promise that drugs could help anxious people feel better, "traditional talk-therapists have lost ground to neurologists and pharmacologists". This is in large part due to the influence of pharmaceutical companies on the medical discipline (Migone 2017).[2] Given the far-reaching influence of the pharmaceutical industry, there is a good chance that the treatment of eco-anxiety in the future will meet the same fate. Third, there are other approaches that can complement psychological therapy and/or medication and help to find inspiration to address eco-anxiety. In addition to the phenomenological perspective briefly mentioned in Chapter 2 (see, e.g., Lafontaine 2022; Mickey 2022; Oele 2024; Vaškovic 2023; Vaškovic and Vičanová 2024), other approaches include Christian, Jewish, and Islamic spiritual traditions, Buddhist, Taoist, and Confucian teachings, ecofeminism, Marxist ecological theory, critical social theory, and Indigenous and

---

[1] This second reason is related to the first one as the medicalization of anxiety (and other mental health conditions) is partly caused by the lack of adequate psychological support for anxious people.

[2] There is also a tendency to medicalize grief, as it exhibits properties that resemble other conditions that are medicalized, such as depression. However, according to Michael Cholbi (2021, 168–175), we should resist efforts to classify grief as a mental disorder and treat it with drugs. The main reason for this is that grief represents a healthy response to the loss of those who play a central role in our understanding of who we are and what matters to us. Grief can generate conditions serious enough to merit medical concern, but it is not a pathology that should be medicalized. Alessandri (2024, 133–149) develops a similar line of reasoning in the case of anxiety. As anxiety often represents a normal reaction to the human condition and current social and political pathologies, we ought to resist the tendency to medicalize this emotion: "The medical light makes the human side of anxiety hard to see and overdiagnosis easy, especially in the context of the financial incentives driving the medical industry complex" (Alessandri 2024, 141). This comparison between grief and anxiety, however, has two limitations. First, while anxiety is an emotion, grief represents a series of affective states, which can include anxiety, but also anger, acceptance, guilt, or fear. Second, as discussed in Chapter 2, anxiety can be pathological in that it can lead to mental health disorders such as phobia, social phobia, panic disorder, GAD, obsessive–compulsive disorder, and PTSD. In contrast, grief is not pathological, and even though there have been discussions about proposing a mental disorder specific to grief, called "complicated grief disorder", this proposal has not been included in the DSM. As Cholbi (2021, 185) writes, "this does not preclude those whose grief leads to depression, anxiety, or other conditions, being eligible for medical assistance. It merely suggests that their grief itself is not the grounds or rationale for their being entitled to such assistance".

First Nations perspectives (see, e.g., Banwell and Eggert 2024; Jamieson and Nadzam 2025; Vaškovic 2023; Wiseman 2021).

This chapter explains why ecological citizenship is a relevant normative framework for facing eco-anxiety. First, it helps to mitigate the undesirable side effects of eco-anxiety by embedding pro-environmental behaviour and ecological attitudes in the political identity of individuals. Second, it helps to harness the beneficial effects of eco-anxiety by making individuals more receptive to the relevant ways of addressing contemporary ecological problems, both at the individual and collective levels. Third, it contributes to the development of a moral and political community of like-minded citizens who face ecological risks head-on and express the core values of carbon sobriety, hope, and courage in their daily actions.

The chapter starts with a definition of the notion of "citizenship" and explains why it has become necessary to articulate national with post-national forms of citizenship, such as cosmopolitan and ecological citizenship. Then, it explains that ecological citizenship is a normative framework that can help with navigating eco-anxiety, both through individual and collective actions. It also links individual mitigation actions with the virtue of carbon sobriety, one of the key values of ecological citizens. The rest of the chapter explains how hope and courage are linked, showing in what sense they also represent core ecological citizenship values that can help us to face eco-anxiety, focusing especially on good hope and radical hope.

From National to Ecological Citizenship

Citizenship refers to a status that arises out of membership of a political community that confers a set of specific and reciprocal rights and correlative duties. Correspondingly, citizenship theory is about the definition of the boundaries and the conditions of membership of distinct political communities. One of the most influential conceptions of citizenship, both in theory and in practice, is that of "national citizenship". According to Sue Donaldson and Will Kymlicka (2011, 55–56), citizenship serves three main functions:

(1) *Nationality*: citizenship is about belonging to a given national community;
(2) *Popular sovereignty*: citizenship is about being included in the sovereign people;

(3) *Democratic political agency*: citizenship is about actively partici-pating in the political decision-making process.

The idea that there is an intrinsic association between citizenship and nationality is widespread in political philosophy, where both nationalists (Kymlicka 1995; Miller 1995) and statists (Rawls 1999b; Blake 2001) argue that duties of distributive justice owed by citizens cannot extend beyond the boundaries of the sovereign, territorial nation states. For all these political theorists, citizenship expresses membership of a polity with territorial boundaries within which citizens can enjoy particular rights and exercise their political agency. In other words, "citizenship, both as a legal status and as an activity, is thought to presuppose the existence of a terri-torially bounded political community, which extends over time and is the focus of a common identity" (Leydet 2017).

The assumption that citizenship and nationality are intrinsically associ-ated has, however, been challenged for a very long time by cosmopolitan theorists. The idea of humans belonging to a single community with a global political structure has been a constant from the Greek Stoics to contemporary political theorists, each age producing its own interpreta-tion (Heater 1996). The core concept of cosmopolitanism is the idea that people are "citizens of the world", that humans belong to a single moral and/or political community by virtue of their common humanity and/or a shared economic, political, and cultural framework. Cosmopolitan theorists use the notion of citizenship to expand individuals' responsibil-ities beyond national borders, with the idea that cosmopolitan citizens have first and foremost a duty to support and further the development of just institutional arrangements at the global level (Archibugi and Held 1995; Held 1995; Linklater 1998, 2007; Tan 2017). They uncouple the notion of citizenship from the notion of nationalism and link it instead to that of cosmopolitanism.

A more recent form of post-national citizenship that has been discussed in green political theory is that of ecological citizenship. The expression "ecological citizenship" was coined in the mid-1990s as a renewed and expanded notion of citizenship that would help humanity deal with global environmental problems, such as anthropogenic mass extinction, climate change, and ozone depletion (Barry 2002; Christoff 1996; Dobson 2003; Dobson and Bell 2006; Dobson and Valencia Sáiz 2005a; Smith 1998). Ecological citizenship is also grounded in the awareness that we share a common world, but it shifts the focus from common humanity and shared institutional schemes to common natural resources and services we use to sustain our lifestyles. The ecological footprint of our everyday actions and the environmental impacts of the policies we contribute to designing create an ecological space that is not limited to the national community: the causes and effects of contemporary ecological issues are spatially and temporally dispersed and involve countless individual and collective agents. Ecological citizenship identifies cross-border flows of pollution, natural resources, and energy that lead to transgressions in planetary boundaries to stress that national boundaries are no longer the only relevant markers of the political community.

Ecological citizenship is based on the core idea that to adequately address global environmental change, we ought to adopt more sustainable lifestyles, that is, lifestyles that are more in line with natural limits (Bourban 2023a, b). A major contribution by green political theorists to citizenship theory is to consider changes in behaviours and in underlying attitudes to be as important to citizenship as political participation in the decision-making process that determines the terms of social cooperation. Their way of conceiving the political community is different from that of statists, nationalists, and cosmopolitans. To the communities of fellow nationals, of world citizens, or of members of a common scheme of domestic or global distributive justice, they add the community of members of a shared ecological space. Since the ecological space is generated by our ecological footprint, and since our ecological footprint is generated by "the metabolistic and material relationship of individual people with their environment" (Dobson 2003, 106), activities of citizenship are no longer limited to the public sphere. This is a major conceptual innovation, since traditional forms of citizenship, in both the liberal and the civic republican traditions, limit citizenship activity to political relationships and participation in decision making in the public arena. As Andrew Dobson and Derek Bell (2006, 7) stress, ecological citizenship therefore "invites us to take a fresh look at a crucial piece of the architecture of citizenship with a view, perhaps, towards recasting the mould in which it has traditionally been formed".

In the ecological model of citizenship, the private realm becomes a site of citizenship activity. Dobson highlights that "private acts have public implications of a citizenship sort", and that ecological citizenship "is *all about* everyday living" (Dobson 2003, 135, 138 – emphasis original). Multiple actions in the private realm, such as household energy use, choice of personal diet, or the size of one's family have public consequences because they contribute to environmental impacts at the collective level. Likewise, John Barry (2006, 26) explains that citizenly behaviour as a consumer is possible: "there are citizenly and sustainability virtues and practices to be found within the private/domestic sphere". One can be a good ecological citizen not only as a voter, an elector, an activist, but also as a consumer, a producer, a parent, or a worker. Since strict separation between the private and the public realm can contribute to obscuring ecological injustices by considering only public activities to be potentially just or unjust, the politicization of the private sphere has become necessary if we are to achieve more just and sustainable societies.

This does not mean that ecological citizenship is not also about political participation. Contributing to the design and implementation of more just and effective environmental policies is also a major concern for ecological citizens. They support social, economic, and political measures by local or national governments, but they also support voluntary changes in behaviours as well as in the attitudes and values that underlie behaviours. Self-interested actions based on economic incentives and disincentives, such as carbon taxes, rubbish taxes, and road-pricing schemes play a role in the development of more sustainable societies. However, sustainability requires voluntary shifts in attitudes at a deep level—deeper than those reached by fiscal measures (Dobson and Valencia Sáiz 2005b; Dobson and Bell 2006; Dobson 2007).

One key objective of ecological citizenship is to provide a broader picture of human motivation than the one provided by an approach focused on self-interested behaviours aligned with incentives and disincentives provided by fiscal measures. To the external or extrinsic motivation to protect the environment procured by legal and economic instruments, ecological citizenship adds internal or intrinsic motivations based on the duties, responsibilities, and virtues of environmentally aware citizens. Virtuous citizens internalize the purpose and value of good environmental practice, thus basing their obedience not only on mere external motivations of price, punishment, or prohibition but on self-imposed duties and autonomous virtuous activities (Connelly 2006). Individual attitudes and behaviours are inevitably influenced by regulation, education, and incentives set by governments, but at the same time, citizens can develop an individuality that is relatively or partially independent from the economic and political structures that inform their attitudes and behaviours (Dobson 2007).

Ecological citizenship is especially applicable to addressing current ecological problems such as climate change. Limiting global warming to well below 2 °C and contributing to efforts to limit the temperature increase to 1.5 °C requires radical social, political, and economic reforms and substantial changes in our lifestyles, behaviours, and habits. At a systemic level, the IPCC (2018, 15) stresses that "Pathways limiting global warming to 1.5 °C with no or limited overshoot would require rapid and far-reaching transitions in energy, land, urban and infrastructure (including transport and buildings), and industrial systems". At an individual level, it highlights that lifestyle changes can also substantially contribute to reducing global GHG emissions, for instance, in the form of "balanced, sustainable healthy diets", "adaptive heating and cooling choices", and "shifts to walking, cycling, shared pooled and public transit", as well as "sustainable consumption" (IPCC 2022, 38). It also stresses that these demand-side mitigation options have strong co-benefits in terms of health, employment, energy security, and equity (IPCC 2022, 44). Deep decarbonization pathways rely on a similar approach with three major scenarios focusing on three emissions reduction measures (van Vuuren et al. 2018): the lifestyle change scenario relies on radical shifts towards more environmentally friendly behaviours such as plant-based diets, changes in transport habits, and reductions in heating and cooling levels; the renewable electricity scenario relies on the substantial expansion of solar and wind technologies, based on their encouraging progress over the last few years; the low-population scenario is based on population policies that would reduce birth rates around the world. One of the best ways to link systemic changes with lifestyle changes is to rely on ecological citizenship, since the relationship between the collective and the individual levels is at the core of this theoretical model. Ecological citizenship aligns with the levels and scale of change required to properly mitigate climate change.

ECOLOGICAL CITIZENSHIP AND ECO-ANXIETY

How can ecological citizenship be relevant to navigating eco-anxiety? On the one hand, ecological citizens are more likely to be eco-anxious since their civic commitment is based on an environmental awareness that is itself based on knowledge of the state of the planet and/or on experienced ecological impacts. It would probably be exaggerated to claim that eco-anxiety is a condition of possibility for ecological citizenship, but it is very likely that many, if not most, ecological citizens have feelings of eco-anxiety because of their knowledge of environmental problems and/or their experience of their impacts. On the other hand, ecological citizens are also more likely to find constructive ways to cope with eco-anxiety, since they perceive citizenship as intrinsically linked with environmental action, both at the individual level of lifestyle choices and at the collective level of policymaking. Even though their knowledge and/or experiences can initially make them more vulnerable to eco-anxiety, they do not necessarily let themselves be paralysed by its psychological burden. The idea here is that although ecological citizenship often goes hand in hand with eco-anxiety, it is worth developing it as a coping strategy to learn to live with this ecological emotion.

Ecological citizenship can help both to address and to harness eco-anxiety. A point briefly discussed in Chapter 2 and that I will develop now is that different forms of environmental action can help address eco-anxiety and its effects on mental well-being. Drawing on the climate anxiety compass developed by Anne van Valkengoed and Linda Steg (2024), it is useful to distinguish between individually oriented and collectively oriented mitigation actions. The first category includes lifestyle changes such as changing diets that can help mitigate mild forms of eco-anxiety; reducing flying and meat consumption can also help address medium forms of eco-anxiety, and avoiding flying altogether may be a mitigation strategy for more severe forms of eco-anxiety (Hickman 2020). The second category includes group actions such as activism and campaigning that help to address both significant and severe forms of eco-anxiety, even if the benefits of these actions may be limited when the effects of eco-anxiety on mental health are detrimental. Since ecological citizenship is about both individual and collective environmental action, it can help to develop effective mitigation strategies for all forms of eco-anxiety.

Actions at the Individual Level

Focusing first on individually oriented mitigation actions, empirical studies have shown that eco-anxiety can encourage environmental action. Reviewing the literature on health implications of eco-anxiety, Boluda-Verdú et al. (2022, 1) stress that "eco-anxiety was associated with determinants of pro-environmental behavior". Eco-anxiety has, for instance, been associated with pro-ecological world views and a green self-identity.

Importantly, eco-anxiety does not automatically lead to environmental action; specific conditions must be met (Van Valkengoed et al. 2023). First, the degree of eco-anxiety matters: while higher levels can be overwhelming and paralysing, moderate levels may be conducive to environmental action. Second, the relation between eco-anxiety and environmental action can be mediated, for instance, through other ecological emotions, such as anger, or through cognitions, such as one's perception of efficacy, capacity, and responsibility. I argue here that ecological citizenship can act as one possible form of mediation in this context: by perceiving themselves as ecological citizens, it is more likely that eco-anxious people will perform environmental actions. In other words, ecological citizenship can strengthen the link between eco-anxiety and environmental actions. The reason for this is that, as stressed above, ecological citizens aim to live more sustainable lifestyles and see this objective as constitutive of their political and moral identity.

The ecological citizen's principal obligation is to have a (more) sustainable ecological footprint.[3] Drawing on the Brundtland report's conception of sustainable development (WCED 1987, 8), Dobson (2003, 119) states, "the ecological citizen will want to ensure that his or her ecological footprint does not compromise or foreclose the ability of others in present and future generations to pursue options important to them". Now, what would a sustainable ecological footprint look like? Dobson does not really develop the practical implications of ecological citizenship and therefore does not really answer this important question. He defines the ecological footprint as follows: "the 'space' of ecological citizenship is created by the metabolistic relationship between individual human beings (and collections of them) and their non-human natural environment as they go about producing and reproducing their daily lives. This is the 'ecological footprint'" (Dobson 2003, 131). But how is it to be measured? Different components of the ecological footprint—waste, pollution, or GHG emissions—may be easier or more difficult to measure. One key component is the *carbon footprint*, which can be measured in terms of individual GHG emissions. What would a sustainable ecological footprint look like in this instance?

Climate change literature puts the sustainable level at approximately 2 tons of CO_2-equivalent (tCO_2e) per person, per year (Fragnière 2016,

---

[3] I draw here on Bourban (2025a).

806; Wynes and Nicholas 2017, 1; UNEP 2020, xxv); some authors put it even lower, at 1.3 tCO_2e (Chancel and Piketty 2015). Most affluent people have a carbon footprint that is several times higher. For instance, the annual per capita consumption-based CO_2 emissions of an average American, Canadian, or Dutch citizen are 15.5, 12.9, and 8.7 tons, respectively (OurWorldInData 2022). To meet the demands of ecological citizenship, affluent citizens therefore need to substantially reduce their GHG emissions. Where exactly the upper threshold should be is a complex question: the level differs from country to country and, within a given country, from locality to locality, based on criteria such as availability and affordability of low-carbon energy or of public transport. It is, however, safe to say that average levels of per capita emissions in developed countries tend to be unsustainable and are therefore above any reasonable interpretation of the upper threshold.

To get closer to a sustainable carbon footprint, ecological citizens undertake as many high-impact actions as reasonably possible, that is, actions that reduce their GHG emissions by at least 0.8 tCO_2e per year (Wynes and Nicholas 2017). In developed countries, there are five major high-impact actions to substantially reduce annual personal emissions (see Fig. 3.1):

(1) Adopting a plant-based diet (0.8 tCO_2e saved per year)
(2) Buying green energy (1.5 tCO_2e saved per year)
(3) Avoiding air travel (1.6 tCO_2e saved per round-trip transatlantic flight)
(4) Living car-free (2.4 tCO_2e saved per year)
(5) Having one less child (58.6 tCO_2e saved per year)

There are three important remarks to make here. First, how effective and how demanding these actions are varies. They also belong to different categories: while actions (1) to (4) can be categorized as consumption-based actions, (5) is a fundamental life choice and therefore requires much more justification.[4] The common ground of all high-impact actions is their requirement that we question the dominant, high-emitting lifestyles of industrial societies.

---

[4] I develop such a justification elsewhere (Bourban 2019; 2025b). I do not focus on this topic here, as this specific—and controversial—high-impact action is not central to my argument.

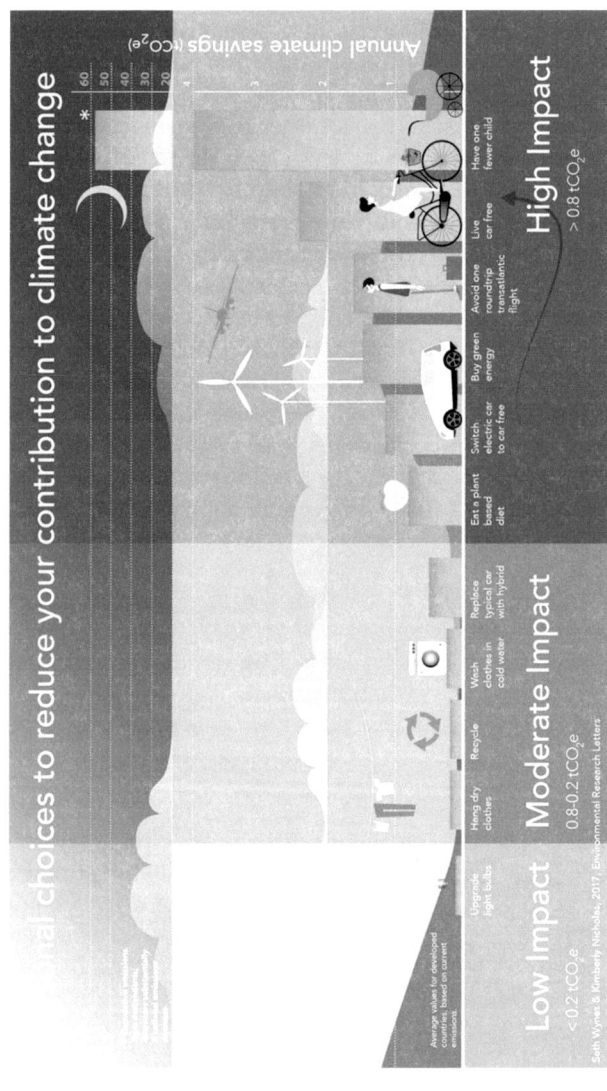

Fig. 3.1 Low-, Moderate-, and High-Impact Actions to Reduce One's Carbon Footprint in Developed Countries. Data from Wynes and Nicholas (2017) (Credit: Catrin Jakobsson)

Second, although adults have the greatest capacity to adopt high-impact actions, children and young people can also contribute to reducing their own carbon footprint and that of their household. Although the choices listed here in terms of transportation and family size are essentially not up to them, they can have an influence on their parents' choices in these domains. Also, in the domain of food consumption, children and young people can adopt a plant-based diet. Their capacity to stick strictly to such a diet is constrained by the possibilities offered by their families and institutions such as schools, but they can also shift the feasibility set here by influencing the options available to them in terms of food consumption. These are relevant examples of ecological citizenship acts.

Third, since there are so many carbon-dependent structures in industrialized countries, living a low-carbon lifestyle can be quite demanding. A general ethical principle to keep in mind here is the Kantian rule "ought implies can": people should not be required to do things that are not in their power (Baatz 2014; Bourban 2023c; Fragnière 2016). The options and possibilities open to individual agents differ considerably, depending on their respective geographical, economic, and social situations. This means that individuals are morally required to reduce their emissions *only to the extent that they are capable of doing so.* The possibility of implementing high-impact actions is partially out of moral agents' control, since they depend, for instance, on the socio-economic classes to which they belong, or on the availability and affordability of renewable energy.

Importantly, this does not mean that only people in developed countries can and should adopt these actions: a rich person in a developing country has more opportunities and responsibility to reduce their carbon footprint than a poor person in a developed country. There are two important reasons for this (Bourban 2022). First, the richer people are, the greater their carbon footprint. According to Oxfam (2015), while the richest 10% of the world's population are responsible for around 50% of global emissions, the poorest 50% are responsible for only around 10% of global emissions.[5] Someone in the richest 10% emits on average *60*

---

[5] Other studies have confirmed this key finding. Chancel and Piketty (2015) found the richest 10% contribute about 45% of global emissions, while the bottom 50% contribute only 13% of global emissions. Similarly, the UNEP (2020) found that while the top 10% of earners are responsible for 48% of global emissions, the bottom 50% of earners only account for 7% of global emissions. The top 1% of earners alone are responsible for 15% of global emissions, which is more than twice the amount emitted by the bottom 50%. More generally, there is a positive correlation between affluence and overall environmental

times more than someone in the poorest 10%; the richest 1% emits *175 times* more than the poorest 10%. Second, since the lifestyles of wealthy citizens are characterized by an abundance of choice, they are in the best position to reduce or avoid consuming high-emitting goods and services that contribute the most to environmental degradation (Wiedmann et al. 2020).

Actions at the Collective Level

Moving now to the category of collectively-oriented mitigation actions, ecological citizens can promote and support collective action against climate change. They can typically do so by helping to design institutions that are better adapted to tackling climate change. In this context, ecological citizens have a wide range of possibilities open to them, such as writing blogs and articles, petitioning their local government, emailing their representatives or executives, organizing and/or attending demonstrations, and donating to or joining NGOs (Cripps 2013). They can also vote green, join civil disobedience actions in response to government failure, and push for more ambitious climate policies (Caney 2014). Here again, adults, children, and young people can all act as ecological citizens, as attending demonstrations, joining an NGO, or putting pressure on local and national representatives are actions that can be taken regardless of whether or not one is of legal voting age. The same applies to legal action against governments, as children and young people are also involved in lawsuits against their governments because of their failure to protect ecosystems, young citizens, and their future (Parker et al. 2022). This participation in activist efforts and litigation cases represents ecological citizenship actions that allow children and young people to reclaim their agency and escape the trap of paralysis.

In the case of individually oriented action, as we saw above, the "ought implies can" rule matters. At the level of collectively oriented action, another intuitive rule is what Simon Caney (2014, 141) calls the "Power/Responsibility Principle", according to which "those with the power to compel or induce or enable others to act in climate-friendly ways have a responsibility to do so". This implies that different actors bear different

---

impact: while the world's top 10% of income earners are responsible for between 25 and 43% of environmental impact, the world's bottom 10% of income earners exert only between 3 and 5% of environmental impact (Teixidó-Figueras et al. 2016).

responsibilities for promoting and supporting collective action against environmental problems. For instance, climate scientists—who are, just like children and young people, more exposed to experiencing eco-anxiety—can play a part in undermining resistance to ambitious climate policies by rebutting factual errors and misleading statements by climate deniers. More generally, researchers can combat misinformation and disinformation on social media and elsewhere through public outreach events, publications, and research projects that promote scientific literacy, strengthen trust in scientific findings, and promote informed decision making based on scientific evidence.[6] To further widen the scope, journalists, poets, novelists, and gifted communicators can promote green lifestyles; lawyers can contribute to climate litigation and help those who engage in lawsuits against states and corporations by providing legal expertise; and engineers can design more sustainable power plants, buildings, and infrastructures. A wide range of actions is available to different kinds of actors, but it is possible for most individuals to contribute, in their own way, to collective efforts to address environmental problems.

This connection between ecological citizenship and the collective level is especially important in the context of eco-anxiety. As we saw in Chapter 2, many eco-anxious people experience feelings of abandonment, isolation, or alienation. This is one of the key findings of the global survey of children and young people's thoughts and feelings about climate change and government responses to climate change. The authors found that feelings of eco-anxiety are correlated with feelings of betrayal and abandonment by governments and adults, linked to perceived inadequate responses to climate change. Adult inaction towards climate change is "experienced as betrayal and abandonment, not just of the individual but of young people and future generations generally" (Hickman et al. 2021, e871). Britt Wray, one of the authors of the study, develops this crucial point in the following way in an interview on the findings of the study:

> I have hundreds of people reach out and tell me that their anxiety is made so much worse by the fact that if they try to talk about the concerns in their circles of friends and family and those people aren't ready to hear their concerns and legitimize them, they can end up feeling many times

---

[6] See, for instance, the EU-funded VERITY (developing scientific Research with ethIcs and integrity) and IANUS (Inspiring and Anchoring TrUst in Science) research projects (European Commission 2023a, b).

worse. It leads to feeling very isolated and alienated in the intensity of these emotions, which makes them very difficult to cope with (cited in Mufarech 2022).

The study on the state of eco-anxiety in France discussed in Chapter 1 made a similar observation (Sutter et al. 2025). It found that people who are strongly eco-anxious become increasingly isolated from those around them, because they increasingly come into conflict with others over environmental issues. It also found that those who are very strongly eco-anxious tend to be ostracized by those around them, who can no longer stand them talking about environmental issues all the time. Strongly and very strongly eco-anxious people therefore experience social isolation, which often stems from a feeling of incomprehension of being out of sync with others (even those who are closest to them) who do not necessarily share the same concern about environmental problems.

This shows that eco-anxiety is fundamentally *relational*: it is linked to the way we relate to other people, both those close to us, such as friends and family members, but also those who are more distant, such as co-citizens and government officials. Eco-anxiety is a subjective emotional state, but the person feeling this emotion is also affected by the discourse and behaviour of people around them. It is an emotional response to both the state of the world and the perception of what others are or are not doing about it, a response to both global environmental problems and the unresponsiveness of individual and collective agents to these problems. It is this discrepancy between this feeling of urgency and the inadequate response or the lack of any response that creates these feelings of abandonment, isolation, and alienation.

Ecological citizenship can help here, as it represents one possible way to reconnect with people who (1) feel concerned about the state of the planet, and (2) are actively doing something about it, both at the individual level of lifestyle changes and the collective level of political decision making. Ecological citizens belong to a shared moral and political community where the object of eco-anxiety, ecological risks, as well as the mitigation of such risks, are at the centre of citizenship activity. This can help someone to combat feelings of alienation and to find purpose by joining forces with like-minded people.

In the same spirit, ecological citizenship can help to harness the more positive side of eco-anxiety discussed in Chapter 2. As we saw, eco-anxiety is not an intrinsically negative emotion; it can also have beneficial effects. Eco-anxiety typically leads to information gathering behaviours, concerned with good or accurate decision making, such as making effective lifestyle changes. It can also push us to connect with other people to create change at the collective level. This is another point explained by Wray (2022, 9), who highlights the revitalizing power of eco-anxiety:

> mourning ecological losses may hold political power for us now, allowing us to gather the conviction to make change while extending a platform for others to join us. Rather than bury our heads in the sand and suppress our discomfort, we can harness and transform the distress we feel into meaningful actions and forms of connection.

If properly channelled, and if felt to a degree that is not too high, eco-anxiety can help us to connect with other people. Since ecological citizenship is both about lifestyle changes and participation in the decision-making process, it can help to harness these positive effects of eco-anxiety and turn them into environmental action rather than letting eco-anxiety push us to paralysis.

Because of its association with feelings of guilt and paralysis, eco-anxiety can be perceived as an obstacle to addressing collective action problems and a source of depoliticization of the climate crisis. This happens, for instance, when eco-anxious people reduce existing sociopolitical pathways to apocalyptic scenarios or when they reduce environmental action exclusively to individual actions and personal behaviour. However, eco-anxiety can also be a source of re-politicization, especially when linked with ecological citizenship. In this context, eco-anxiety can push people to look for more information on climate change, to cooperate with other people to address it in a collective fashion, and to put pressure on governments to pursue more ambitious environmental policies. Ecological citizenship can help eco-anxious people to reconnect with their political subjectivity. It can help turn eco-anxiety into a source of political action, or, as Joe Davidson (2024, 437) puts it, a "productive starting point for political mobilisation and coalition-building".

In addition to pushing local and national government to adopt more ambitious environmental policies, ecological citizens can also support efforts at the collective level to address eco-anxiety through adequate public health policies. These include (positive) awareness policies on eco-anxiety, investments in the research on eco-anxiety (especially on how to address its most radical forms), and investments in training for mental health professionals to make them qualified to help eco-anxious people. This last point is especially important, as psychologists and psychiatrists cannot be expected to develop knowledge on how to diagnose and treat eco-anxious people on their own; they need support at the collective level, for instance, in the form of free training sessions on how to deal with eco-anxiety in order to prevent its most extreme forms, typically by developing a network of coordinators trained in eco-anxiety issues (Sutter et al. 2025).

Let me finish this section with two important observations. First, although ecological citizenship and environmental action are valuable, there is no guarantee that they can fully address eco-anxiety. A first limitation is that individual actions need to be aggregated at the collective level to make a discernible difference over the course of climate change. Another limitation is that activism can also negatively affect people's mental health, potentially leading to activist fatigue or even burnout (Van Valkengoed 2023). Although it is not a silver bullet for addressing eco-anxiety, ecological citizenship still represents a promising normative framework for learning to live with eco-anxiety by transforming it into a mode of action and a political stance.

Second, the fact that eco-anxiety can be associated with environmental action does not imply that eco-anxiety should be cultivated, encouraged, or, worse, inculcated. It is important not to neglect the beneficial aspects of eco-anxiety, but it is equally important to stress its detrimental and even potentially pathological effects on mental health. The main source of motivation to address environmental problems should not be eco-anxiety; it should rather be environmental awareness, ecological citizenship, and environmental values. Eco-anxiety is neither a consistent nor a necessary source of motivation; values and norms are more robust and stable (Van Valkengoed 2023). As eco-anxiety is associated with distress and suffering that can affect day-to-day functioning, mitigating it should be a priority for both researchers and practitioners.

Carbon Sobriety

What are the key values of the ecological citizen? In the remainder of this chapter, I investigate three core values that are especially relevant in the context of eco-anxiety: (carbon) sobriety, (good and radical) hope, and (environmental) courage.

Values can take the form of *virtues*, which represent a commitment to living well. A virtue is an acquired and stable disposition to do good by being good. They are moral values lived out in action, expressed through our behaviour. The most important is not so much achieving some measurable good result as becoming a good person whose behaviour expresses key ethical commitments. In the eudemonic approach to virtue ethics, virtues are not only about living well; they are also about living happily. As Allen Thompson (2010, 50) explains, "a virtue is a character trait that a person needs to live well in the sense that possession of the virtues is constitutive of human flourishing". He gives the example of frugality, without developing it, but which is very close to the virtue of temperance supported by Dale Jamieson (2007, 181), which is about "self-restraint and moderation" and leads to "reducing consumption", or the virtue of simplicity advocated by Joshua Gambrel and Philip Cafaro (2009), an attitude towards material goods that typically includes decreased consumption.

In the same vein, I conceive *carbon sobriety* as a green virtue that represents an alternative to consumerism and another way to live well with less.[7] The etymology of sobriety (*sōphrosýnē* in Greek, *sobrietas* in Latin) refers to a self-imposed form of moderation, a sensible measure. It is the opposite of disproportionate behaviour and excessive consumption. Sobriety has a long history of conceptual development, starting with the Stoics and the Epicureans; today, sobriety can be applied to the environmental impact of our lifestyles. As Bruno Villalba (2023, 631) stresses, sobriety can be conceived as a "mindset" centred on the idea that "we must keep our needs, desires and behaviours proportional to what the planet is able to cope with". Carbon sobriety represents a key virtue that helps ecological citizens fulfil their duty to adopt a sustainable carbon footprint.[8] It relies on a moral motivation to practise self-limitation, which guides ecological citizens' everyday living. It is not a matter of

---

[7] Following the French literature on environmental theory and politics (Arnsperger and Bourg 2014; Pelluchon 2017; Villalba and Semal 2018), I used in a previous publication the notion of "energy sobriety" (*sobriété énergétique*) (Bourban 2022). I prefer now the wider notion of "carbon sobriety", as it includes emitting activities that are not predominantly based on energy, such as emissions for food production and consumption. I develop a fuller account of carbon sobriety based on ecological limitarianism in Bourban (2025).

[8] I do not rely here on a religious or theological conception of virtue, but on a secular approach to virtue ethics focused on the dispositions or character traits that make us "good" ecological citizens.

heroic morality or asceticism, but rather an acquired and stable disposition to enjoy consuming less, to lead a flourishing life by living more soberly when it comes to emitting activities, especially high-emitting activities. It is not merely a practice, but a state of character that informs our everyday actions, a fundamental choice of lifestyle that is more sustainable, more consistent with natural limits. Carbon sobriety leads to specific courses of action, but it is above all about who we want to be and the moral motives that drive us to act—or not—in a certain way.

Carbon sobriety is a green virtue in the sense that it helps promote both human and non-human flourishing. It is based on the realization that humans do not flourish in an ecological vacuum: our individual and collective well-being depends on ecological resources and services provided by a flourishing natural world. This is why individual, social, and ecological flourishing are interrelated. As Cafaro (2015) highlights,

> our flourishing and nature's flourishing are intertwined. It is no accident that the same actions and the same personality traits typically help us to be good neighbors and citizens and good environmentalists. The same ecosystems, in many cases, facilitate the flourishing of human and nonhuman beings; pollution and declining ecosystem health harm both people and other organisms.

An ecological citizen whose behaviour reflects the virtue of carbon sobriety undertakes as many high-impact actions as possible to reduce their carbon footprint. As discussed above, such actions typically cover changes in diets (adopting a vegan diet), transportation habits (avoiding long-distance air travel or travel by car), energy source (buying exclusively green energy from one's energy provider), and family size (having a small family).

Carbon sobriety is a key trait of the ecological citizen who wishes to reduce their carbon footprint. A more sustainable lifestyle allows exploration of non- or less-consumptive ways of achieving well-being. Carbon sobriety is not about self-sacrifice, but about having more desirable relationships with oneself, others, and the natural world. This is why it can be conceived as a green virtue: it is about both human and ecological flourishing, and the relation between the two.

Let me make two important remarks before moving on to the next value. First, virtues such as carbon sobriety motivate us to act regardless of what our actual impact on the course of global climate change is. Virtues are not only a matter of good results; they are first and foremost related to the intention and the capacities of the person who acts (Bourban and Broussois 2020). The objective is to reduce one's carbon footprint but also to develop new lifestyles that are self-rewarding and contribute to happiness and well-being. Virtues are character traits that motivate us to act regardless of the behaviour of others. As Jamieson (2014, 186) explains, they "give us the resiliency to live meaningful lives even when our actions are not reciprocated". Virtues sustain patterns of behaviour whatever others do or do not do. A person should act virtuously, even if it has no discernible effect in the world, because they are especially responsible for what they do, no matter what others do.

Second, virtuous actions can also encourage others to adopt greener lifestyles and give up on the culture of consumerism. This is due to the communicative power of individual actions: efforts to reduce personal emissions can have an amplifying effect and lead other individuals to do the same, thereby contributing to more substantial emission reductions. The decision to change one's lifestyle might inspire others to make similar changes and contribute to a collective shift. Marion Hourdequin (2010, 457) explains this phenomenon the following way: "individual consumer decisions, personal conversations about such decisions, and similar small-scale, local actions may turn out to be important catalysts for emerging collective agreements". Ecological citizens can strongly encourage this phenomenon of amplifying effect by framing this new form of living well through greener lifestyles as appealing so as to influence other people to reduce their own carbon footprint. Abandoning the rhetoric of self-sacrifice and explaining the different ways in which more virtuous lifestyles can be self-rewarding and contribute to happiness and well-being is a powerful tool to push other people to go green (Prinzing 2023).

HOPE

A second value that intuitively comes to mind to counteract anxiety is that of hope (Bourban 2024). As Darrel Moellendorf (2022, xii) writes, "Hope is a tonic against resignation and debilitating anxiety". In discussions on ecological emotions, hope has been linked with optimism and empowerment, but it is not fully clear what it means. As Panu Pihkala (2022, 17) stresses, "there is a strong need to inquire further about the actual meanings of 'hope' for various people and scholars". This section explains what hope means and why it is also a valuable disposition for facing eco-anxiety.

Hope, Anxiety, and Despair

To find a clearer picture of what hope involves, I draw here on Dominic Roser's account, which relies on three elements: desire, belief, and emphasis (Roser 2020). According to this account, hope applies to an object that is (1) desired, (2) believed to be possible but that remains uncertain, and (3) characterized by a certain mental emphasis that makes the desire and the belief of the hoper significant and stable. There is a mental emphasis on the possibility of a positive outcome, but that does not mean that hope only applies to outcomes that are likely. As Roser (2020, 68) explains, "I can believe X to have a low probability (condition 2.) but can still desire X (condition 1.) and psychologically rally around X (condition 3.)". He illustrates this with the case of the 1.5 °C climate limit: even if the probability of limiting temperature increase to 1.5 °C is low, this goal remains possible and desirable and can therefore be an object of hope.

This is the key difference between hope and optimism: while optimism is marked by a high degree of confidence in success, hope is characterized by a non-confident belief in the possibility of the desired outcome. Hope takes place in a context of doubt: the object of hope need not be very likely or even likely at all; it only needs to be reasonably possible. This point is also stressed by Moellendorf (2022, 31): "hope is not confidence that things will turn out as we want. It is not optimism. Instead, hope arises when the prospects are uncertain or even doubtful". Even if the outcome is considered unlikely or doubtful, it is not unreasonable to hope for it (see also Williston 2012).

Hope can lead to wishful thinking, disappointment, and distraction, but these disadvantages can be compensated by the motivational and hedonic benefits that come with it: a hopeful disposition can represent a source of personal motivation to overcome collective action problems and can lead to a high degree of perseverance vis-à-vis the object of hope (Roser 2020; Williston 2012). This important point is highlighted by other climate justice scholars, such as Catriona McKinnon (2014, 45), who stresses that "Hope can increase the probability that a person's agency achieves its purpose, and so can galvanise the person's will as it aims at this purpose", and Moellendorf (2022, 32), who explains that

"Hoping for a future state of affairs supports motivation to act to achieve the good and avert the ill". The object of hope here can be the goal of maintaining global warming below 1.5 °C or well below 2 °C, but it can also be more indeterminate, for instance, reaching a minimal form of climate justice or even avoiding the worst climate injustices.

There are therefore good reasons to cultivate hope to mitigate eco-anxiety, but at the same time, one needs to treat hope carefully. The reason for this is that hope is an ambiguous emotion. It can be both a source of and a hindrance to motivation to act. This has to do with both the possible sources and effects of hope. First, hope can be illusory when it is based on wrong beliefs or ignorance, such as the belief that ecological problems will resolve without the need for human intervention or without the need for radical lifestyle changes. Second, hope can be short-lived and can hold people back from preparing for and accomplishing relevant actions and can instead lead to deception and even harm (Oele 2024). This is also recognized by Roser (2020, 77), who stresses that "Dwelling on the imagined achievement instead of working towards it, hinders rather than spurs action". Dwelling on the prospect of the hoped-for object is a form of distraction from the planning necessary for working towards it. Third, although hope can represent a relief from a harsh reality and the negative emotions related to it, it can also lead to a denial of the seriousness of the climate problem and therefore a denial of any responsibility to take action as a way to cope with this reality (Ojala 2017).

Hope can therefore also be a hindrance to successfully taking actions that achieve the hoped-for object. The main reason for this is that hope is not, contrary to a commonly shared conception, the opposite of anxiety. The opposite of hope is rather despair, a feeling typically generated by a judgement that the object of hope is impossible or extremely improbable because of the way the world is (McKinnon 2014). Despair involves what Anthony Steinbock (2007, 171) calls the "loss of the ground of hope as such". Hope and anxiety are actually strongly linked as they are both based on *doubt*: just like the object of anxiety, the object of hope is characterized by uncertainty. The important difference is that while we strive for the object of hope, we wish to avoid the realization of the object of anxiety, but that does not make hope and anxiety mutually exclusive. On the contrary, it is difficult to have one without the other, as it is common to hope that the object of anxiety will not come to fruition, just as it is common to be anxious about not attaining one's object of hope.

This is why, as Spinoza (2018 III, 50, Scholium, 132) observed, "there is no hope without fear nor fear without hope". To hope is to fear disappointment; to fear is to hope for reassurance. Importantly, Spinoza's conception of fear is close to anxiety understood as an emotional response to problematic uncertainty: "Fear [*metus*] is uncertain sadness arising from the idea of something in the future or in the past about whose outcome we are to some extent in doubt" (Spinoza 2018 III, Definition 13, 146). As we saw in Chapter 2, future orientation and doubt are key features of anxiety. Note how close Spinoza's definition of hope is to his definition of fear: "Hope [*spes*] is uncertain joy arising from the idea of something in the future or in the past about whose outcome we are to some extent in doubt" (Spinoza 2018 III, Definition 12, 146). There is therefore a strong proximity between anxiety and hope, both conceptually (when it comes to the definition of these two related emotions) and phenomenologically (when it comes to the experience of these emotions, what it feels like to be anxious and hopeful).

The element of doubt is key to understanding the proximity and relation between anxiety and hope. If doubt is removed, in the case of hope, we are left with assurance (or confidence) [*securitas*], a "joy arising from the idea of something in the future or in the past about which the cause to have doubts has been taken away"; in the case of fear, we are left with despair [*desperatio*], a "sadness arising from the idea of something in the future or in the past about which the cause to have doubts about it has been taken away" (Spinoza 2018 III, Definitions 14 and 15, 147).

While eco-anxiety and hope are closely linked, pushed to too high a degree, eco-anxiety can also lead to *eco-despair*. At that stage, the eco-anxious person believes that a just sustainability transition has become impossible or is extremely improbable, and that the darkest scenarios such as the Hothouse Earth pathway are virtually unavoidable. The uncertainty about the realization of a wide range of possible futures gives way to a quasi-certainty of the realization of a very narrow set of threatening futures. For instance, the hope of keeping global temperatures below 1.5 °C, or even 2 °C, is replaced by the strong belief that global warming of 4 °C or 5 °C by 2100 is almost unavoidable, with catastrophic consequences for human and non-human beings. For eco-desperate people, environmental action no longer makes sense, aside perhaps from radical adaptation actions to prepare for a very dangerous world. Eco-despair can lead to depression, anxiety disorders, and suicidal thoughts.

Grounds to Remain Hopeful

Looking at existing climate scenarios, we might think that there are good reasons to give in to eco-despair. If we are to comply with the overall goal of the Paris Agreement to limit global warming to "well below 2 °C" while "pursuing efforts to limit the temperature increase to 1.5 °C" (UNFCCC 2015, art. 2.1), the remaining carbon budget is small and rapidly shrinking. The IPCC (2022, 10) warns that the remaining carbon budget is no more than 500 $GtCO_2$ for a 50% probability of limiting global warming to 1.5 °C and no more than 1150 $GtCO_2$ for a 67% probability of limiting global warming to 2 °C. To get an estimate of how little this represents, cumulative net CO_2 emissions for the 2010–2019 decade would exhaust about four-fifths of the totality of the remaining carbon budget for a 50% probability of limiting global warming to 1.5 °C, and about one-third of the remaining budget for a 67% probability of limiting global warming to 2 °C (IPCC 2022, 10). Still in the last IPCC Assessment Report, we can read that global emissions pathways consistent with implemented policies will lead to total warming of 3.2 °C by the end of the century, with a range of 2.2 °C to 3.5 °C (IPCC 2023, 22). This provides a very bleak picture of our future.

At the same time, the IPCC also stresses that the 3.2 °C scenario is just one scenario among many others. For instance, it shows that modelled pathways that have a higher than 50% chance of limiting warming to 1.5 °C, or a higher than 67% chance of limiting warming to 2 °C, with no or limited overshoot of the global carbon budget, are possible. However, these scenarios will be socially and economically very demanding, as they "involve rapid and deep and, in most cases, immediate greenhouse gas emissions reductions in all sectors this decade" (IPCC 2023, 20). Scenarios with limited overshoot also rely on carbon dioxide removal (CDR), a climate engineering measure that removes CO_2 from the atmosphere to store it in geological, terrestrial, or ocean reservoirs or in products. CDR can rely on highly engineered negative emissions techniques (NETs), such as bioenergy with carbon capture and storage (BECCS), direct air capture with carbon storage (DACCS), enhanced weathering, and ocean fertilization. It can also rely on non-technological measures or options that rely much less on technologies, such as ecosystem restoration, alternative agricultural practices, afforestation, and reforestation. Most scenarios that keep global warming well below 2 °C and aim for a temperature increase of no more than 1.5 °C imply a substantial carbon budget overshoot compensated by hundreds of gigatons of negative emissions over the second half of the century (Riahi et al. 2021, 1063).

Large-scale CDR would allow us to expand our very limited global carbon budget so that the overshoot of the remaining budget during the century would be compatible with achieving stringent warming targets by 2100. Given current emissions trajectories and the limited results of subnational, national, and international mitigation policies, it seems like a long-awaited deus ex machina. The idea that hundreds of gigatons of negative emissions will be available in the coming decades is, however, problematic as there is no guarantee that this will be economically, technologically, politically, or biophysically possible – or, indeed ethically desirable (Bourban 2026). On the one hand, the remaining global carbon budget is so small at present and projected GHG emissions are so high that some form of CDR is very likely to be required to avoid dangerous climate change. On the other hand, one must be careful not to become over-reliant on a technical measure that only features a few prototypes and projects and whose capacity to be scaled up might prove limited for economical, technological, political, or biophysical reasons (Smith et al. 2016, 2023). This would, for instance, apply to IPCC scenarios requiring the construction of up to 16,000 BECCS plants to extract huge amounts of CO_2 out of the atmosphere (Lenzi et al. 2018, 304).

The reason why there are grounds to remain hopeful without giving in to eco-despair is that there is a substantial empirical literature that shows that it is still possible for a portfolio of mitigation measures to limit global warming to well below 2 °C and contribute to efforts to limit the temperature increase to 1.5 °C with no or limited carbon budget overshoot and only small-scale CDR (see, e.g., Falk et al. 2019, 106; Hansen et al. 2017, 595–596; Höhne et al. 2017, 20–21; Rockström et al. 2017, 1271; Strefler et al. 2018, 2–3). In other words, there is still a safe operating space for humanity, but choosing this future would require radical social, political, and economic reforms, and substantial changes in our lifestyles, behaviours, and habits. This is why ecological citizenship is such an important model to promote and develop: it keeps pathways that limit and even avoid global temperature overshoot within reach. Such pathways are associated with higher upfront investment and near-term mitigation costs.[9]

---

[9] A GDP reduction of 0.5–4.8% in scenarios that keep warming below 1.5 °C with no or limited overshoot and of 0.1–1.6% in scenarios that limit warming to 2 °C with no

However, these higher costs would be fully compensated in the second half of the twenty-first century: "avoiding overshoot would be associated with economic gains in the long term (even without considering benefits of avoided climate impacts)" (Riahi et al. 2021, 1067).

This gives good reason to favour eco-hope over eco-despair. However, there is also good reason to be anxious about what the future holds. First, many existing scenarios, such as the Hothouse Earth pathway are frightening and threatening. Second, even if we follow a deep decarbonization pathway, this might still raise plenty of social, economic, and environmental problems. As stressed in Chapter 2, the object of eco-anxiety is not only the risks raised by environmental issues but also the risks raised by the measures adopted to address them, such as issues of social justice and environmental justice raised by the extraction of rare-earth metals and conflict minerals required for renewable energy technologies. And even if we manage to maintain global temperatures well below 2 °C, other environmental problems such as biodiversity loss and pollution will continue to worsen if we do not address them with complementary environmental policies and actions too. Third, whatever the pathway we choose to follow, we will have to face more frequent and severe climate impacts, even if we manage to maintain global warming to 1.5 °C: "Climate-related risks to health, livelihoods, food security, water supply, human security, and economic growth are projected to increase with global warming of 1.5 °C and increase further with 2 °C" (IPCC 2018, 9). Fourth, "hope-makers" (Moellendorf 2022, 30–34), that is, facts, theories, visions, actions that give reasons to be hopeful, such as the upsurge in climate activism around the world, are threatened by hope-underminers, such as the rise of far-right politics in Europe and the US, as well as the problems of political polarization and misinformation about climate change encouraged by social media (Lenzi 2023). These hope-underminers can reduce the prospects of a mass mobilization for a more just climate future. They can also increase the risk of more people turning to fascism as an alternative to democracy when ecological citizenship fails to deliver the hoped-for change in existing national and international institutions. Since eco-anxiety and eco-hope go hand in hand, there is limited risk of eco-anxiety motivating people to turn to anti-democratic ideas and action; however, there is a reasonable risk

or limited overshoot, compared to end-of-century scenarios with overshoot (Riahi et al. 2021, 1065).

of eco-pessimism and eco-despair encouraging the development of fascist ideologies if hope-underminers were to overtake hope-makers. This is an additional reason to combat both eco-despair and hope-underminers and support and strengthen hope-making initiatives.

The future remains open, but, importantly, it will not remain so indefinitely. As climate change worsens, the window of opportunity to implement minimally socially, economically, and environmentally disruptive measures gradually narrows. The choices we make today and in the close future will greatly influence which options are available and those that will become unavailable for young people and future generations. As Steffen et al. (2018, p. 8253) argue:

> social and technological trends and decisions occurring over the next decade or two could significantly influence the trajectory of the Earth System for tens to hundreds of thousands of years and potentially lead to conditions that resemble planetary states that were last seen several millions of years ago, conditions that would be inhospitable to current human societies and to many other contemporary species.

It is important to distinguish here between *irreversibility* and *unpredictability* (Guillaume and Petit 2017). On the one hand, the existence of planetary boundaries in the Earth system implies that there are non-negotiable ecological limits and that, once critical tipping points are crossed, there is no coming back (Steffen et al. 2015). In this sense, the future is indeed gradually narrowing. Positive feedback processes may play a more important role than assumed in some IPCC scenarios, "limiting the range of potential future trajectories and potentially eliminating the possibility of the intermediate trajectories" (Steffen et al., 2018, p. 8253). If we greatly overshoot the carbon budget, there are significant risks that the planetary threshold of 2 °C will be crossed during the century. At this point, the Earth system is likely to follow an "essentially irreversible pathway driven by intrinsic biogeophysical feedbacks" that could lead to a cascade of tipping points pushing the Earth irreversibly into a Hothouse Earth pathway (Steffen et al., 2018, p. 8254). So there is indeed irreversibility in the climate system, which explains why we are in a "state of planetary emergency" (Lenton et al. 2019, p. 595). Insufficient emissions reductions lead not to temporary but to irreversible effects in the climate system.

On the other hand, there is still unpredictability, because we can still avoid crossing the critical threshold that would trigger a cascade of tipping points in the climate system. Steffen et al. (2018, p. 8257) also stress that "Widespread, rapid, and fundamental transformations will likely be required to reduce the risk of crossing the threshold and locking in the Hothouse Earth pathway; these include changes in behaviour, technology and innovation, governance, and values". In addition to institutional and social innovations at the global level, "changes in demographics, consumption, behaviour, attitudes, education, institutions, and socially embedded technologies" are all important to maximize the chances of achieving a 'Stabilized Earth Pathway'" (Steffen et al., 2018, p. 8257). In other words, multiple different scenarios are still possible at this point, including deep decarbonization pathways. The future is still open, albeit not for long.

The future should therefore not be perceived as something fully open or fully closed; it should rather be perceived as a countdown. This countdown is well illustrated by the idea of the carbon budget. The longer we wait to substantially reduce global emissions, the narrower the future becomes. We can either slow down or even stop the countdown by avoiding a large overshoot, or we can accelerate it or even reduce it to zero by missing the opportunity to quickly and drastically reduce global emissions. We, therefore, find ourselves at a very strategic moment in the history of humankind with an unprecedented responsibility. Humanity is now facing the need for critical decisions and actions, and the reason we can still make decisions is that the future is still open, even though our window of opportunity to avoid catastrophic climate change is closing rapidly. Given this picture, hope is still warranted, but anxiety is difficult to avoid. Being anxiously hopeful seems like the most adapted emotional response to our planetary predicament.

The Value of Hope

Hope is a valuable disposition for learning to live with eco-anxiety. I see four main reasons for this: it encourages a realistic outlook on the future, thereby distinguishing itself from optimism and pessimism; it encourages a certain form of solidarity; it affects the probability of success of achieving the hoped-for outcome; and it provides hedonic benefits (Kadlac 2015; Moellendorf 2022; Roser 2020). Let us develop these four points in turn.

First, hope can lead to having a more realistic outlook on the future. While the optimistic and the pessimistic person base their outlook on the future mainly on beliefs about that future, the hopeful person focuses instead mostly on the probabilities relating to that future. Hope is *evidence-sensitive*. The hopeful person is especially attentive to the evidence concerning whether certain future events will occur, which provides them with a more appropriate assessment of future possibilities than that of the optimistic or the pessimistic person. The eco-desperate person overestimates the probability and/or severity of ecological risks and dangers and becomes confident that the future will be catastrophic, thereby missing positive features in their present circumstances. By doing so, they tend to isolate themselves, as many people will avoid being around someone who refuses to acknowledge the possibility of success and who believes that all is doom and gloom. In contrast, the eco-hopeful person acknowledges the inherent uncertainty in determining between the many possible scenarios that may obtain in the future, and this keeps them motivated to pursue the outcome they desire.

As Adam Kadlac (2015, 345) stresses, "Being hopeful, then, allows us to acknowledge, and even indulge, our desires about the future while being fully cognizant of the uncertainty of that future". This attitude towards the future allows the eco-hopeful person to use their knowledge about ecological risks to estimate the probability that a given state of affairs will obtain, while still maintaining a desire that things will turn out the way they want them to. They are very much aware of the gap between their desires and reality, but this does not prevent them from remaining hopeful, as uncertainty can cut both ways: it can increase or reduce the gap. Eco-hope is an antidote to eco-fatalism.

While the eco-hopeful person is not pessimistic, they are not optimistic either. A high degree of confidence in success would lead them to ignore manifest obstacles to the achievement of the object of their desire. This can also be an obstacle to motivation to act, as overconfidence can lead to the belief that not much additional effort is needed to steer policy development and technological innovation in the right direction.[10]

---

[10] A recent empirical study found that an optimistic bias in updating beliefs about climate change leads to low engagement in pro-environmental behaviour, "possibly because selectively integrating good news over bad news reduces the perceived urgency to take action" (Kube et al. 2025, 1). Not sufficiently considering undesirable information may diminish one's sense of urgency and individual responsibility. More optimistic people

Techno-optimism, excessive confidence in the capacity of technological artefacts to address social, political, and environmental problems, is a good example of the problematic aspects of optimism. It typically leads to techno-solutionism, the misplaced belief that cheap techno-fixes can help "solve" climate change without much effort from individuals.

The second reason is that hope affects the probability of achieving the hoped-for outcome. A perceived capacity to find ways to achieve desired goals and to motivate oneself to implement them leads to better outcomes in many domains, including in physical and mental health (Snyder 2002). The third reason is that hoping has hedonic benefits: it provides pleasure and solace and brightens up everyday life, thereby balancing the stress and worry that comes with eco-anxiety. It feels good to hope, and this helps, in part, to live with eco-anxious feelings. Hope, especially under the form of good hope, is a valuable emotion that should be cultivated. It is a disposition of the ecological citizen that helps them to deal with problematic uncertainties raised by environmental problems.

The fourth reason is that the hopeful person aims to create or join a community of people jointly committed to the same project to increase the chances that their desire will come to fruition. They want to face ecological risks with others and work with them towards reducing the likelihood and severity of those risks. They want to promote a just and sustainable society at the collective level and join people who desire the same kind of future and are motivated to work towards bringing that future about. This is why Kadlac (2015, 350) highlights that hope encourages a certain kind of solidarity with others: "one reason to think that solidarity is an important value is because it recognizes the numerous ways in which we are all vulnerable to an uncertain future". By stressing our vulnerability in the face of contingencies that are beyond our control, solidarity has a strong social value that is directly linked to the social value of hope, a disposition that has at its heart a recognition of our uncertain future. The hopeful person is committed to solidarity as they are looking for others who share their hopes and uncertainties and are willing to fight for their desires. This brings us to the topic of good hope.

---

are less likely to engage in environmental action as they believe that the situation is under control or is being managed by others.

Good Hope

Hope can help couple environmental awareness with environmental action, despite lingering uncertainties, but only if the right kind of hope is pursued. In other words, to avoid the potentially problematic effects of hope, such as denial, distraction, and wishful thinking, it is important to cultivate constructive forms of hope, that is, hope that helps eco-anxious people to constructively deal with their emotions.[11] In this context, good hope and radical hope are especially apposite.

As Victoria McGeer (2004, 122) explains, "maintaining hope for most of us requires a somewhat responsive social world—a world of at least some particular others who recognize and support the meaning and value we give to our hopeful efforts". Good hope is intrinsically *social*: in that sense, it can help eco-anxious people, such as children and young people but also environmental scientists and the many researchers, educators, and students who work on environmental sciences, to feel less isolated, alienated, or ostracized. As stressed above, eco-anxiety has a strong relational aspect: one of the reasons why it might feel so overwhelming and discouraging is that many eco-anxious people experience their social world as unresponsive to their concerns. Engaging in good hope can help eco-anxious people to find a community of like-minded people, who are both concerned about the state of the world and willing to do something about it. Hoping well has an interpersonal dimension: "it depends on finding— or making—a community in which individual hopers can experience the benefits of peer scaffolding" (McGeer 2004, 123). In this sense, acting hopefully as ecological citizens means not only undertaking high-impact environmental actions, but also finding ways to socialize these actions by building a community and practices around them and thereby making them more impactful.

Ecological citizens do not hope that their individual environmental actions will have any impact on the course of climate change and other environmental issues; they are aware that this would just be wishful thinking, as individual actions, taken separately, are too small to make any real difference to global environmental changes. However, as members of a moral and political community of environmentally aware and active citizens, they are hopeful that their common efforts, taken together, can

---

[11] For recommendations for educators on how to transform the concept of "constructive hope" into practice, see Ojala (2017).

lead to sustainable changes. As members of a shared moral and political community, they live in a social world that is at least partially receptive to their commitments to the environment and where eco-anxiety is not a taboo but a widely shared emotion. They contribute to creating a world that is supportive of their hopes. Dale Jamieson (2011, 36) conceives this community as a "morally motivated global citizens' movement that acts as a highly committed political interest group". This is how he describes it:

> Such a movement would stigmatize coal, meat eating, trophy houses, over-heating and overcooling, large living spaces, and private automobiles. It would celebrate living lightly with dignity and elegance, relying on nature's own energy, rediscovering food and the pleasures of eating, the joys of living with nature and other people, and the satisfaction of effective political activism. I think (and hope) that we may be witnessing the birth of such a movement.

Relying on a community of ecological citizens who share similar hopes significantly helps to maintain and strengthen environmental agency in a context of transgression of planetary boundaries. This community is not a guarantee that our hopes will come to fruition. As McGeer (2004, 124) stresses, "there is no ultimate protection from the disappointment of unrealized hopes": the hopes that ecological citizens have of a just ecological transformation of their societies may be frustrated by realities that will leave them disappointed and unfulfilled. However, hoping together with responsive others is a protection from eco-despair that makes it possible to transform eco-anxiety into motivation to mitigate ecological risks rather than letting these risks become a source of isolation and paralysis. Hope is partly a skill that can be trained, by learning to adjust it to one's power to act, abilities, and resources. It is a disposition that can be developed over time.[12]

---

[12] The significance of hope is also influenced by culture. Some cultures demand positive narratives or progress and growth, and hope is indexed to them. In contrast, other societies favour continuity of tradition, and hope is related to that context. This means that what hope implies can (and probably will) change from one culture to another, even if the core of good hope, its social and relational features, remains unchanged.

Radical Hope

A second form of hope that can help with learning to live with eco-anxiety is that of radical hope. Radical hope was originally discussed by Jonathan Lear as a way of thinking about how hope could persist for the Crow people when faced with the end of their way of life and cultural collapse (Lear 2008). He focuses especially on the tribe's leader, Plenty Coups, who brought the Crow people through by maintaining hope in a meaningful future even when they could not conceive of what that future might look like. Radical hope is hope in the face of profound uncertainty and the collapse of familiar cultural structures.

This form of hope may also be appropriate when societies start to face ecological collapse. Saying that there is a serious risk of social, economic, and political collapse caused by environmental factors is no longer a catastrophist statement based on exaggerations of the available scientific information. Take, for instance, the following statement by Earth system scientists:

> Currently, the Earth System is on a Hothouse Earth pathway driven by human emissions of greenhouse gases and biosphere degradation toward a planetary threshold at ~ 2 °C [...], beyond which the system follows an essentially irreversible pathway driven by intrinsic biogeophysical feedbacks. (Steffen et al. 2018, 8254)

In some scenarios, such as the Hothouse Earth pathway, ecological collapse may take place at the global level. In other scenarios, it may take place at a regional or local level, for instance, inhabitants of small island developing states (SIDS), such as Tuvalu or Kiribati, being forced to leave their homeland because of sea level rise and other climate impacts, thereby becoming climate exiles (Byravan and Rajan 2010). There is a growing literature on the topic of ecological collapse. The emerging transdisciplinary field of "collapsology" addresses the possible causes of collapse of civilization as we know it, as well as the possible ways to live in a post-collapsed world (Servigne and Stevens 2015).

Radical hope can help to address the psychological challenges raised by the possibility of ecological collapse. It represents an openness to a profoundly uncertain future, when the cultural or existential world is at risk of dissolving. Instead of letting eco-anxiety spiral into pessimism or despair, radical hope can help us focus on the possibility of meaning beyond the loss of the world as we know it. It is not only an awareness of the possibility of ecological loss; it is also a recognition of the possibility of transformation (new values and new forms of relationships with nature), even when the future is threatening. It helps to live with eco-anxiety without being overwhelmed by it.

Radical hope is not a form of pessimism. As explained above, eco-pessimism is based on an overestimation of the severity or the likelihood of ecological risks and dangers. It is a confident expectation of future ecological catastrophe, which can lead to missing genuinely positive features of present circumstances. Hope, including radical hope, avoids this way of looking at things by acknowledging the uncertainty of whether a given situation will obtain in the future, which allows individuals to pursue the desired outcome. Typically, eco-pessimistic people strongly believe that ecological collapse will take place in the near future. Collapsology can be a source not only of eco-anxiety but also of eco-pessimism and even eco-despair. It highlights the possibility of ecological collapse, stating for instance that "The times we live in are clearly marked by the spectre of collapse", but it also supports more radical statements, such as the idea that "We are therefore probably experiencing the last gasps of the engine of our industrial civilisation before its extinction" (Servigne and Stevens 2015, 32, 63). This doom-and-gloom approach can lead to a collapse of hope and can just as easily lead to a counterproductive form of eco-despair characterized by an apolitical conception of solidarity and a survivalist tone (Charbonnier 2019). If we believe that civilizational extinction is nigh, or that there is very little chance that humanity or much else will survive, eco-anxiety gives way to eco-despair. In this case, the behavioural response of risk mitigation is replaced by a response focused mostly, if not exclusively, on extreme forms of adaptation, such as preparation for catastrophic climate change or ecological collapse, taking the form of stockpiling food, ammunition, and other supplies. In contrast with ecological citizens, so-called preppers are more self-centred and individualistic and do not try to build a moral and political community to mitigate ecological risks.

The implication is that radical hope is not only hope in the face of ecological collapse—which is indeed one possible scenario, but among many other possible scenarios, such as deep decarbonization pathways. It is also a form of hope that persists even when the familiar framework that previously helped make sense of the world starts to collapse. Here, the object of collapse is no longer some form of ecological stability; it is dominant ethical and cultural frameworks. For instance, climate change is likely to cause significant changes in the culture of consumerism, understood as "a set of attitudes and values leading people always to high levels of consumption and orienting them to find meaning and satisfaction in life largely through the practices of purchasing new consumer goods" (Thompson 2010, 44). This culture of consumerism greatly influences our cultural and moral landscape by shaping how we relate to nature and value the non-human world. Scenarios such as the Hothouse Earth pathway force us to anticipate conditions of scarcity in which the cultural framework of consumerism can no longer stand. Radical hope can serve to conceive new ethical values and new lifestyles that will allow us to envisage living on a planet that may be very different from the one on which human civilization has developed and to which current forms of life on Earth have adapted. Mitigating the damages of climate change therefore requires personal commitment to the struggle against pervasive consumerist dispositions: "radical hope is manifest as a kind of commitment, in restraint and a resolve against taking too much satisfaction in the consumptive modes to which we are accustomed" (Thompson 2010, 51–52).

What radical hope can push us to preserve in the face of eco-anxiety is our ability to flourish as moral agents. While ecological collapse represents a threat to human flourishing, radical hope can help us turn our energy to actively creating a new world that will not be (too) hostile to human flourishing. The radically hopeful person finds effective routes to their goals and sees their way through or around obstacles. As Byron Williston (2012, 183 – emphasis original) stresses, "finding a way to flourish in the teeth of the climate crisis *requires* working for meaningful political change, acting in newly courageous ways, and looking hard for alternative models of sustainable living".

What is the object of radical hope in this context? I see two relevant and complementary options.[13] The first one would be to maintain the conditions of possibility for justice in the future. Drawing on Hume,

---

[13] There are many additional objects radical hope can have here, such as hope for what Will Steffen et al. (2018) would call a "Stabilized Earth pathway" in which humanity plays an active planetary stewardship role, or hope for what Anna Tsing (2021) would call "living in the ruins" by shifting from a mindset of control and mastery over nature to one of collaborative survival with each other and the non-human world and finding possibility amidst ecological devastation.

Rawls (1999a, 109) explained that to be possible, justice needs specific circumstances, which correspond to the "normal conditions under which human cooperation is both possible and necessary". Among the objective circumstances that make human cooperation both possible and necessary, the condition of moderate scarcity is particularly crucial: "Natural and other resources are not so abundant that schemes of cooperation become superfluous, nor are conditions so harsh that fruitful ventures must inevitably break down" (Rawls 1999a, 110). In a society or a world characterized by general scarcity, such as the Hothouse Earth pathway, the background conditions that make human cooperation and the pursuit of justice possible are no longer given.[14] The idea of scarcity is key here: the Hothouse Earth pathway is characterized by the scarcity of natural resources and services, such as water, arable land, and carbon sinks. However, the most important idea to understand what is at stake here is that of disruption of global natural systems caused by flows of matter and energy underlying our economic activities. The crossing of planetary boundaries leads not only to problems of scarcity but also (and mostly) to the disruption of the conditions of possibility for a flourishing human life on this planet.

A second object of radical hope would be to maintain the meta-capability of sustainable ecological capacity. As Breena Holland (2008) stresses, environmental sustainability can be conceived as a "meta-capability", that is, a condition of possibility for the realization of all the other capabilities. All the central human functional capabilities outlined by Martha Nussbaum (2006, 76–77) in her list of basic capabilities depend on the natural environment. For instance, to live a human life of normal length or not die prematurely, we need ecological systems that provide food, fresh water, the ingredients for medicine, and energy. To have good

---

[14] This is an interpretation of Rawls's account of the circumstances of justice. Rawls himself had very little concern about environmental issues and failed to see the relevance of the environmental precondition claim. He mentioned environmental problems near the end of *A Theory of Justice*, but only to explain that concerns about such problems are not appropriately political but rather embodied in metaphysical doctrines: "A correct conception of our relations to animals and to nature would seem to depend upon a theory of the natural order and our place in it. One of the tasks of metaphysics is to work out a view of the world which is suited for this purpose" (Rawls 1999a, 448). Environmental problems such as "the destruction of a whole species" can indeed "be a great evil", but they are beyond the "limits of a theory of justice" (Rawls 1999a, 448). At no point does Rawls consider that environmental stability can represent a condition of possibility for just institutions or play a crucial role in enjoying core human interests.

health and adequate nourishment, we need ecological systems such as soil, water, and specific atmospheric conditions—and so on. The meta-capability of sustainable ecological capacity can be defined as follows: "being able *to live one's life in the context of ecological conditions that can provide environmental resources and services that enable the current genera-tion's range of capabilities; to have these conditions now and in the future*" (Holland 2008, 324 – emphasis original). Since functioning ecological systems are necessary for a flourishing human life, they are necessary for the exercise of any human capability.

COURAGE

Just as there is a direct link between anxiety and hope, there is also a strong link between hope and courage. Continuing to hope in the face of the real possibility of disappointment or even harm requires substantial courage. This link is especially clear in the case of radical hope as it can be conceived as a form of courage appropriate to a culture in crisis, a courage in the face of cultural collapse (Thompson 2010). When one is facing the risks that attend the collapse of one's culture, courage becomes a necessary component of hope (Lear 2008). But it is also possible to state, more generally, that the courageous person is also usually a hopeful person, as they approach the future and its inherent uncertainty with their eyes wide open (Kadlac 2015).

Broadly speaking, courage is a willingness to risk sacrificing something important to defend a worthy cause. The courageous person incurs the risk of sacrificing something of importance to them, not because they do not fear the consequences of their actions, but because they accept these consequences. It is not the absence of fear but the ability to act in favour of a moral cause despite that fear. This is why courage helps us to translate ethical values into action: the moral dissociation between ethical commitments and actual behaviour is often caused by a lack of willingness to take the risks that result from the realization of these commitments. Courage involves acting in accordance with one's ethical convictions in spite of the risks that entails: "moral courage involves facing the particular fears and dangers arising from the possibility that one will be punished (broadly speaking) for taking a moral stand" (Pianalto 2012, 167).

The coping strategies of individual and collective climate action require not only carbon sobriety and hope; they also require a fair amount of courage. One important reason for being eco-anxious is the threat to environmental value posed by current ecological issues. This moral value can be intrinsic (when the natural world or some of its parts have value in their own right) or instrumental (when nature or its parts have value only insofar as they benefit human beings or their interests), but in any case, it is threatened and needs to be protected.

From this perspective, as Rachel Fredericks (2014) highlights, courage can also be conceived as a green virtue. It is a stand in support of the recognition of value in the environment (as a whole or in the natural entities in it, instrumentally or intrinsically) despite the risk of punishment arising from the protection of this value. Taking action to protect environmental value, in the form of both collective and individual mitigation action, does require courage. In the case of collective mitigation action, environmental activists face physical dangers during arrest and in jail, but they also face non-physical dangers, such as being harassed on social media and becoming exhausted and even suffering from burnout. In the case of individual mitigation action, there are also social risks that one has to face as a result of consumption choices or lifestyle changes, such as risks of criticism and ostracism because of a commitment to the vegan lifestyle. Lifestyle changes that express carbon sobriety are courageous because (1) they are made to protect something morally important, that is, environmental value, and (2) they are carried out despite potential punishments. They are not all equally courageous, as the degree of courage depends on factors such as the number, severity, diversity, and probability of the risks involved, but since they all share these two characteristics, they all display some form of courage.

Courage is, therefore, not only the disposition to act despite the risks arising from ecological threats; it is also the willingness to face the risks arising from individual and collective actions to address these ecological threats. This is why it is an important complement to hope to turn eco-anxiety into action. While a hopeful disposition is a source of perseverance that helps us get closer to the object of hope and leads to a realistic outlook on how to obtain it, a courageous disposition gives the moral agent a willingness to act despite being afraid of the dangers they are facing. The two dispositions are actually complementary. Maintaining hope in our state of planetary emergency requires a certain form of courage: we live in a dangerous world where six out of nine planetary boundaries have already been crossed, where we are at risk of triggering critical tipping points in the climate system, and where the Earth System is currently on a Hothouse Earth pathway. This is why climate hope should be coupled with climate courage (Astola 2025; Wiseman 2022). Carbon sobriety, hope, and courage represent the key ethical values of ecological citizens. A hopeful disposition is a condition for living soberly and courageously, as it represents a shield against eco-despair, which can lead to paralysis, and eco-optimism, which can lead to complacency. Good hope and radical hope can motivate eco-anxious people to face the ecological risks with their eyes wide open and to adhere to carbon sobriety to address those risks despite the consequences they might face. While good hope and radical hope help people react to ecological threats individually and collectively in an appropriate way, courage helps them face the risks of addressing those threats.

REFERENCES

Alessandri, Mariana. 2024. *Night Vision: Seeing Ourselves through Dark Moods*. Princeton University Press.

APA. 2013. 'Recognition of Psychotherapy Effectiveness'. *Psychotherapy* 50 (1): 102–109. https://doi.org/10.1037/a0030276.

Archibugi, Daniele, and David Held. 1995. *Cosmopolitan Democracy: An Agenda for a New World Order*. Polity Press.

Arnsperger, Christian, and Dominique Bourg. 2014. 'Sobriété Volontaire et Involontaire'. *Futuribles* 403: 43–57.

Astola, Mandi. 2025. 'Climate Anxiety and Climate Courage'. *Think* 24 (71): 33–39.

Baatz, Christian. 2014. 'Climate Change and Individual Duties to Reduce GHG Emissions'. *Ethics, Policy & Environment: A Journal of Philosophy and Geography* 17 (1): 1–19. https://doi.org/10.1080/21550085.2014.885406.

Banwell, Nicola, and Nadja Eggert. 2024. 'Rethinking Ecoanxiety through Environmental Moral Distress: An Ethics Reflection'. *The Journal of Climate Change and Health* 15 (January): 100283. https://doi.org/10.1016/j.joc lim.2023.100283.

Barry, John. 2002. 'Vulnerability and Virtue: Democracy, Dependency, and Ecological Stewardship'. In *Democracy and the Claims of Nature*, edited by B. A. Minteer and B. P. Taylor. Rowman & Littlefield.

Barry, John. 2006. 'Resistance Is Fertile: From Environmental to Sustainability Citizenship'. In *Environmental Citizenship*, edited by Andrew Dobson and Derek Bell. The MIT Press.

Blake, Michael. 2001. 'Distributive Justice, State Coercion, and Autonomy'. *Philosophy & Public Affairs* 30 (3): 257–296.

Boluda-Verdú, Inmaculada, Marina Senent-Valero, Mariola Casas-Escolano, Alicia Matijasevich, and María Pastor-Valero. 2022. 'Fear for the Future: Eco-Anxiety and Health Implications, a Systematic Review'. *Journal of Environmental Psychology* 84 (December): 101904. https://doi.org/10.1016/j.jenvp.2022.101904.

Bourban, Michel. 2019. 'Croissance démographique et changement climatique: repenser nos politiques dans le cadre des limites planétaires'. *La Pensée écologique* 3 (1): 19–37. https://doi.org/10.3917/lpe.003.0019.

Bourban, Michel. 2022. 'Ethics, Energy Transition, and Ecological Citizenship'. In *Comprehensive Renewable Energy*. Elsevier. https://doi.org/10.1016/B978-0-12-819727-1.00030-3

Bourban, Michel. 2023a. 'Ecological Citizenship'. In *Handbook of the Anthropocene*, edited by Nathanaël Wallenhorst and Christoph Wulf. Springer International Publishing. https://doi.org/10.1007/978-3-031-25910-4_168.

Bourban, Michel. 2023b. 'Ecological Cosmopolitan Citizenship'. *Future Humanities* 1 (1): 1–21. https://doi.org/10.1002/fhu2.5.

Bourban, Michel. 2023c. 'Mitigation Duties'. In *Handbook of the Philosophy of Climate Change*, edited by Gianfranco Pellegrino and Marcello Di Paola. Handbooks in Philosophy. Springer International Publishing. https://doi.org/10.1007/978-3-031-07002-0_52.

Bourban, Michel. 2024. 'Eco-Anxiety: A Philosophical Approach'. In *Anxiety Culture: The New Global State of Human Affairs*, edited by John Allegrante, Ulrich Hoinkes, Michael Schapira, and Karen Struve. Johns Hopkins University Press.

Bourban, Michel. 2025a. 'Rethinking Climate Justice: Toward Ecological Limitarianism'. *Ethics, Policy & Environment*, November 10, 1–18. https://doi.org/10.1080/21550085.2025.2574219.

Bourban, Michel. 2025b. 'Taking Population Seriously in the IPAT Equation: Ethical and Political Implications of Planetary Boundaries'. In *Sufficiency: From Growth and Overshoot to Enoughness*, edited by Toni Ruuska and Tina Nyfors. Brill.

Bourban, Michel. 2026. 'Overshoot and Recover? On the Problem of Substitution between Negative Emissions and Emissions Reductions'. *Environmental Values*: 1–22. https://doi.org/10.1177/09632719261421916

Bourban, Michel, and Lisa Broussois. 2020. 'The Most Good We Can Do or the Best Person We Can Be?' *Ethics, Policy & Environment* 23 (2): 159–179. https://doi.org/10.1080/21550085.2020.1848175.

Byravan, Sujatha, and Sudhir Chella Rajan. 2010. 'The Ethical Implications of Sea-Level Rise Due to Climate Change'. *Ethics & International Affairs* 24 (3): 239–260. https://doi.org/10.1111/j.1747-7093.2010.00266.x.

Cafaro, Philip. 2015. 'Environmental Virtue Ethics'. In *The Routledge Companion to Virtue Ethics*, edited by Lorraine Besser-Jones and Michael Slote. Routledge.

Caney, Simon. 2014. 'Two Kinds of Climate Justice: Avoiding Harm and Sharing Burdens'. *Journal of Political Philosophy* 22 (2): 125–149. https://doi.org/10.1111/jopp.12030.

Chancel, Lucas, and Thomas Piketty. 2015. *Carbon and Inequality: From Kyoto to Paris*. Paris School of Economics.

Charbonnier, Pierre. 2019. 'Splendeurs et misères de la collapsologie. Les impensés du survivalisme de gauche'. *Revue Du Crieur* 13 (2): 88–95. https://doi.org/10.3917/crieu.013.0088.

Cholbi, Michael. 2021. *Grief: A Philosophical Guide*. Princeton University Press.

Christoff, Peter. 1996. 'Ecological Citizens and Ecologically Guided Democracy''. In *Democracy and Green Political Thought: Sustainability, Rights and Citizenship*, edited by Brian Doherty and Marius de Geus. Routledge.

Connelly, James. 2006. 'The Virtues of Environmental Citizenship'. In *Environmental Citizenship*, edited by Andrew Dobson and Derek Bell. The MIT Press.

Cosh, Suzanne M., Rosie Ryan, Kaii Fallander, et al. 2024. 'The Relationship between Climate Change and Mental Health: A Systematic Review of the Association between Eco-Anxiety, Psychological Distress, and Symptoms of Major Affective Disorders'. *BMC Psychiatry* 24 (1): 833. https://doi.org/10.1186/s12888-024-06274-1.

Cripps, Elizabeth. 2013. *Climate Change and the Moral Agent: Individual Duties in an Interdependent World*. Oxford University Press.

Davidson, Joe Pl. 2024. 'The Politics of Eco-Anxiety: Anthropocene Dread from Depoliticisation to Repoliticisation'. *The Anthropocene Review* 11 (2): 427–41. https://doi.org/10.1177/20530196231211854

Dobson, Andrew. 2003. *Citizenship and the Environment*. Oxford University Press.

Dobson, Andrew. 2007. 'Environmental Citizenship: Towards Sustainable Development'. *Sustainable Development* 15 (5): 276–285. https://doi.org/10.1002/sd.344.

Dobson, Andrew, and Derek Bell. 2006. *Environmental Citizenship*. MIT press.

Dobson, Andrew, and Ángel Valencia Sáiz. 2005a. *Citizenship, Environment, Economy*. Routledge.

Dobson, Andrew, and Ángel Valencia Sáiz. 2005b. 'Introduction'. *Environmental Politics* 14 (2): 157–62. https://doi.org/10.1080/09644010500054822

Donaldson, Sue, and Will Kymlicka. 2011. *Zoopolis: A Political Theory of Animal Rights*. Oxford University Press.

European Commission. 2023a. 'deVEloping Scientific Research with ethIcs and inTegritY'. *CORDIS—EU Research Results.* https://cordis.europa.eu/project/id/101058623.

European Commission. 2023b. 'INspiring and ANchoring TrUst in Science'. *CORDIS—EU Research Results.* https://cordis.europa.eu/project/id/101058158.

Falk, J., O. Gaffney, A. K. Bhowmik, et al. 2019. *Exponential Roadmap 1.5.* Future Earth Sweden.

Fragnière, Augustin. 2016. 'Climate Change and Individual Duties'. *Wires Climate Change* 7 (6): 798–814. https://doi.org/10.1002/wcc.422.

Fredericks, Rachel. 2014. 'Courage as an Environmental Virtue': *Environmental Ethics* 36 (3): 339–55. https://doi.org/10.5840/enviroethics201436334.

Gambrel, Joshua Colt, and Philip Cafaro. 2009. 'The Virtue of Simplicity'. *Journal of Agricultural and Environmental Ethics* 23 (1): 85–108. https://doi.org/10.1007/s10806-009-9187-0.

Guillaume, Bertrand, and Victor Petit. 2017. 'Fermeture Des Futurs, Ouverture de La Démocratie'. In *La Démocratie Face Aux Enjeux Environnementaux. La Transition Écologique*, edited by Charles-Yves Zarka. Mimésis.

Hansen, J., M. Sato, P. Kharecha, et al. 2017. 'Young People's Burden: Requirement of Negative CO_2 Emissions'. *Earth System Dynamics* 8 (3): 577–616. https://doi.org/10.5194/esd-8-577-2017.

Heater, Derek. 1996. *World Citizenship and Government.* Palgrave Macmillan.

Held, David. 1995. *Democracy and the Global Order: From the Modern State to Cosmopolitan Governance.* Polity Press.

Hickman, Caroline. 2020. 'We Need to (Find a Way to) Talk about ... Eco-Anxiety'. *Journal of Social Work Practice* 34 (4): 411–424. https://doi.org/10.1080/02650533.2020.1844166.

Hickman, Caroline, Elizabeth Marks, Panu Pihkala, et al. 2021. 'Climate Anxiety in Children and Young People and Their Beliefs about Government Responses to Climate Change: A Global Survey'. *The Lancet Planetary Health* 5 (12): e863–e873. https://doi.org/10.1016/S2542-5196(21)00278-3.

Hoffman, Stefan G., and Jasper A. J. Smits. 2008. 'Cognitive-Behavioral Therapy for Adult Anxiety Disorders: A Meta-Analysis of Randomized Placebo-Controlled Trials'. *The Journal of Clinical Psychiatry* 69 (4): 621–632. https://doi.org/10.4088/JCP.v69n0415.

Höhne, Niklas, Takeshi Kuramochi, Carsten Warnecke, et al. 2017. 'The Paris Agreement: Resolving the Inconsistency between Global Goals and National Contributions'. *Climate Policy* 17 (1): 16–32. https://doi.org/10.1080/14693062.2016.1218320.

Holland, Breena. 2008. 'Justice and the Environment in Nussbaum's "Capabilities Approach"': Why Sustainable Ecological Capacity Is a Meta-Capability'.

Political Research Quarterly 61 (2): 319–332. https://doi.org/10.1177/106
5912907306471.

Hourdequin, Marion. 2010. 'Climate, Collective Action and Individual Ethical Obligations'. *Environmental Values* 19 (4): 443–64. http://www.jstor.org/stable/25764267.

IPCC. 2018. *Global Warming of 1.5°C. An IPCC Special Report on the Impacts of Global Warming of 1.5°C above Pre-Industrial Levels and Related Global Greenhouse Gas Emission Pathways, in the Context of Strengthening the Global Response to the Threat of Climate Change, Sustainable Development, and Efforts to Eradicate Poverty.* Edited by V. Masson-Delmotte, P. Zhai, H. O. Pörtner, et al. World Meteorological Organization. http://www.ipcc.ch/rep ort/sr15/.

IPCC. 2022. 'Summary for Policymakers'. In *Climate Change 2022: Mitigation of Climate Change. Contribution of Working Group III to the Sixth Assessment Report of the Intergovernmental Panel on Climate Change*, edited by P. R. Shukla, J. Skea, R. Slade, et al. Cambridge University Press.

IPCC. 2023. 'Summary for Policymakers'. In *Synthesis Report of the IPCC Sixth Assessment Report*, edited by H. Lee, K. Calvin, D. Dasgupta, et al. Cambridge University Press.

Jamieson, Dale. 2007. 'When Utilitarians Should Be Virtue Theorists'. *Utilitas* 19 (2): 160–183. https://doi.org/10.1017/S0953820807002452.

Jamieson, Dale. 2011. 'Energy, Ethics, and the Transformation of Nature'. In *The Ethics of Global Climate Change*, 1st edn, edited by Denis G. Arnold. Cambridge University Press. https://doi.org/10.1017/CBO978051173229 4.002.

Jamieson, Dale. 2014. *Reason in a Dark Time: Why the Struggle Against Climate Change Failed -- and What It Means for Our Future.* Oxford University Press. https://doi.org/10.1093/acprof:oso/9780199337668.001.0001

Jamieson, Dale, and Bonnie Nadzam. 2025. 'The Case for Spiritual Resilience'. *Carleton College Voice.* https://www.carleton.edu/voice/stories/the-case-for-spiritual-resilience/.

Kadlac, Adam. 2015. 'The Virtue of Hope'. *Ethical Theory and Moral Practice* 18 (2): 337–354. https://doi.org/10.1007/s10677-014-9521-0.

Kube, Tobias, Jasmin Huhn, and Claudia Menzel. 2025. 'Optimistic Bias in Updating Beliefs about Climate Change Longitudinally Predicts Low Pro-environmental Behaviour'. *British Journal of Social Psychology* 64 (3): e12905. https://doi.org/10.1111/bjso.12905.

Kymlicka, Will. 1995. *Multicultural Citizenship. A Liberal Theory of Minority Rights.* Clarendon Press.

Lafontaine, Simon. 2022. 'Anxiety and the Re-Figuration of Action: Living in a Crisis-Shaped Present'. In *Eco-Anxiety and Planetary Hope*, edited by Douglas

A. Vakoch and Sam Mickey. Springer International Publishing. https://doi.org/10.1007/978-3-031-08431-7_4.

Lattie, Emily G., Colleen Stiles-Shields, and Andrea K. Graham. 2022. 'An Overview of and Recommendations for More Accessible Digital Mental Health Services'. *Nature Reviews Psychology* 1 (2): 87–100. https://doi.org/10.1038/s44159-021-00003-1.

Lear, Jonathan. 2008. *Radical Hope: Ethics in the Face of Cultural Devastation*. Harvard University Press. https://doi.org/10.4159/9780674040021.

Lenton, Timothy M., Johan Rockström, Owen Gaffney, et al. 2019. ʼClimate tipping points—too risky to bet against'. *Nature* 575 (7784): 592–595. https://doi.org/10.1038/d41586-019-03595-0.

Lenzi, Dominic. 2023. 'Hope, Pessimism, and the Shape of a Just Climate Future'. *Ethics & International Affairs* 37 (3): 344–361. https://doi.org/10.1017/S0892679423000254.

Lenzi, Dominic, William F. Lamb, Jérôme. Hilaire, Martin Kowarsch, and Jan C. Minx. 2018. 'Don't Deploy Negative Emissions Technologies without Ethical Analysis'. *Nature* 561 (7723): 303–305. https://doi.org/10.1038/d41586-018-06695-5.

Leydet, Dominique. 2017. 'Citizenship'. In *The Stanford Encyclopedia of Philosophy*, edited by Edward N. Zalta.

Linklater, Andrew. 1998. 'Cosmopolitan Citizenship'. *Citizenship Studies* 2 (1): 23–41. https://doi.org/10.1080/13621029808420668.

Linklater, Andrew. 2007. *Critical Theory and World Politics: Citizenship, Sovereignty and Humanity*. Routledge.

Manzoni, G. M., F. Pagnini, G. Castelnuovo, and E. Molinari. 2008. 'Relaxation Training for Anxiety: A Ten-Years Systematic Review with Meta-Analysis'. *BMC Psychiatry* 8 (June): 41. https://doi.org/10.1186/1471-244x-8-41.

McGeer, Victoria. 2004. 'The Art of Good Hope'. *The ANNALS of the American Academy of Political and Social Science* 592 (1): 100–127. https://doi.org/10.1177/0002716203261781.

McKinnon, Catriona. 2014. 'Climate Change: Against Despair'. *Ethics & the Environment* 19 (1): 31–48.

Mickey, Sam. 2022. 'Atmospheres of Anxiety: Doing Nothing in an Ecological Emergency'. In *Eco-Anxiety and Planetary Hope*, edited by Douglas A. Vakoch and Sam Mickey. Springer International Publishing. https://doi.org/10.1007/978-3-031-08431-7_3.

Migone, Paolo. 2017. 'The Influence of Pharmaceutical Companies'. *Research in Psychotherapy: Psychopathology, Process and Outcome* 20 (2). https://doi.org/10.4081/ripppo.2017.276.

Miller, David. 1995. *On Nationality*. Clarendon Press.

Moellendorf, Darrel. 2022. *Mobilizing Hope: Climate Change and Global Poverty*. 1st edn. Oxford University Press. https://doi.org/10.1093/oso/9780190875619.001.0001.

Mufarech, Antonia. 2022. 'Your Crushing Anxiety About the Climate Crisis Is Normal'. *Smithsonian Magazine*, May 18. https://www.smithsonianmag.com/science-nature/how-to-deal-with-the-anxiety-caused-by-the-climate-crisis-180980093/.

Nussbaum, Martha C. 2006. *Frontiers of Justice: Disability, Nationality, Species Membership*. The Belknap Press of Harvard University Press.

Oele, Marjolein. 2024. 'Anxiety, Grief, and Trust in Times of Climate Change: A Phenomenology of Affective Constellations and Future Transformations in and beyond the Anthropocene'. *Comparative and Continental Philosophy*, June 3, 1–20. https://doi.org/10.1080/17570638.2024.2361409.

Ojala, Maria. 2017. 'Hope and Anticipation in Education for a Sustainable Future'. *Futures* 94 (November): 76–84. https://doi.org/10.1016/j.futures.2016.10.004.

OurWorldInData. 2022. 'Per Capita Consumption-Based CO_2 Emissions, 2020'.

Oxfam. 2015. *Extreme Carbon Inequality*. Oxfam International.

Parker, Larissa, Juliette Mestre, Sébastien Jodoin, and Margarentha Wewerinke-Singh. 2022. 'When the Kids Put Climate Change on Trial: Youth-Focused Rights-Based Climate Litigation around the World'. *Journal of Human Rights and the Environment* 13 (1). https://doi.org/10.4337/jhre.2022.01.03.

Pelluchon, Corine. 2017. 'L'éthique des vertus : Une condition pour opérer la transition environnementale'. *La Pensée écologique* 1 (1): 1–36. https://doi.org/10.3917/lpe.001.0101.

Pianalto, Matthew. 2012. 'Moral Courage and Facing Others'. *International Journal of Philosophical Studies* 20 (2): 165–184. https://doi.org/10.1080/09672559.2012.668308.

Pihkala, Panu. 2022. 'Toward a Taxonomy of Climate Emotions'. *Frontiers in Climate* 3 (January): 738154. https://doi.org/10.3389/fclim.2021.738154.

Prinzing, Michael. 2023. 'Going Green Is Good for You: Why We Need to Change the Way We Think about Pro-Environmental Behavior'. *Ethics, Policy & Environment* 20 (1): 1–18. https://doi.org/10.1080/21550085.2020.1848192.

Rawls, John. 1999a. *A Theory of Justice*. Rev. ed. Belknap Press of Harvard University Press.

Rawls, John. 1999b. *The Law of Peoples with 'The Idea of Public Reason Revisited.'* Harvard University Press.

Riahi, Keywan, Christoph Bertram, Daniel Huppmann, et al. 2021. 'Cost and Attainability of Meeting Stringent Climate Targets without Overshoot'. *Nature Climate Change* 11 (12): 1063–1069. https://doi.org/10.1038/s41558-021-01215-2.

Rockström, Johan, Owen Gaffney, Joeri Rogelj, Malte Meinshausen, Nebojsa Nakicenovic, and Hans Joachim Schellnhuber. 2017. 'A Roadmap for Rapid Decarbonization'. *Science* 355 (6331): 1269–1271. https://doi.org/10. 1126/science.aah3443.

Roser, Dominic. 2020. 'The Case for Climate Hope'. In *Edition Moderne Postmoderne*, 1st edn, edited by Nejma Tamoudi, Simon Faets, and Michael Reder. Transcript Verlag. https://doi.org/10.14361/9783839449875-004.

Servigne, Pablo, and Raphaël Stevens. 2015. *Comment Tout Peut s'éffondrer. Petit Manuel de Collapsologie à l'usage Des Générations Présentes*. Seuil.

Smith, Mark J. 1998. *Ecologism: Towards Ecological Citizenship*. Open University Press.

Smith, Pete, Steven J. Davis, Felix Creutzig, et al. 2016. 'Biophysical and Economic Limits to Negative CO2 Emissions'. *Nature Climate Change* 6 (1): 42–50. https://doi.org/10.1038/nclimate2870.

Smith, S. M., O. Geden, G. F. Nemet, et al. 2023. *The State of Carbon Dioxide Removal - 1st Edition*.

Snyder, C. R. 2002. 'Hope Theory: Rainbows in the Mind'. *Psychological Inquiry* 13 (4): 249–275. https://doi.org/10.1207/S15327965PLI1304_01.

Spinoza. 2018. *Ethics: Proved in Geometrical Order*. Edited by Matthew J. Kisner. Translated by Michael Silverthorne. Cambridge University Press.

Steffen, Will, Wendy Broadgate, Lisa Deutsch, Owen Gaffney, Cornelia Ludwig. 2015. 'The trajectory of the Anthropocene: The Great Acceleration'. *The Anthropocene Review* 2(1): 81–98. https://doi.org/10.1177/205301961456 4785.

Steffen, Will, Johan Rockström, Katherine Richardson, et al. 2018. 'Trajectories of the Earth System in the Anthropocene'. *Proceedings of the National Academy of Sciences* 115 (33): 8252–8259. https://doi.org/10.1073/pnas. 1810141115.

Steinbock, Anthony J. 2007. 'The Phenomenology of Despair'. *International Journal of Philosophical Studies* 15 (3): 435–451. https://doi.org/10.1080/ 09672550701445431.

Strefler, Jessica, Nico Bauer, Elmar Kriegler, Alexander Popp, Anastasis Giannousakis, and Ottmar Edenhofer. 2018. 'Between Scylla and Charybdis: Delayed Mitigation Narrows the Passage between Large-Scale CDR and High Costs'. *Environmental Research Letters* 13 (4): 044015. https://doi.org/10. 1088/1748-9326/aab2ba.

Sutter, Pierre-Eric, Jean-Luc Bernaud, and Léonie Messmer. 2025. *Éco-Anxiété En France (Étude 2025)*. ADEME. https://librairie.ademe.fr/societe-et-politi ques-publiques/8137-eco-anxiete-en-france.html.

Tan, Kok-Chor. 2017. 'Cosmopolitan Citizenship'. In *The Oxford Handbook of Citizenship*, edited by Ayelet Shachar, Rainer Bauböck, Irene Bloemraad, and Maarten Vink. Oxford University Press. https://doi.org/10.1093/oxf ordhb/9780198805854.013.30.

Teixidó-Figueras, Jordi, Julia K. Steinberger, Fridolin Krausmann, et al. 2016. 'International Inequality of Environmental Pressures: Decomposition and Comparative Analysis'. *Ecological Indicators* 62 (March): 163–173. https://doi.org/10.1016/j.ecolind.2015.11.041.

Thompson, Allen. 2010. 'Radical Hope for Living Well in a Warmer World'. *Journal of Agricultural and Environmental Ethics* 23 (1–2): 43–59. https://doi.org/10.1007/s10806-009-9185-2.

Tsing, Anna Lowenhaupt. 2021. *The Mushroom at the End of the World: On the Possibility of Life in Capitalist Ruins*. Princeton University Press.

UNEP. 2020. *Emissions Gap Report 2020*. United Nations Environment Programme.

UNFCCC. 2015. *Adoption of the Paris Agreement. Decision 1/CP.21. Document FCCC/CP/2015/10/Add.1*.

Valkengoed, Van, M. Anne. 2023. 'Climate Anxiety Is Not a Mental Health Problem. But We Should Still Treat It as One'. *Bulletin of the Atomic Scientists* 79 (6): 385–387. https://doi.org/10.1080/00963402.2023.2266942.

Valkengoed, Van, M. Anne, and Linda Steg. 2024. 'The Climate Anxiety Compass: A Framework to Map the Solution Space for Coping with Climate Anxiety'. *Dialogues on Climate Change* 1 (1): 39–48. https://doi.org/10.1177/29768659241293226.

Valkengoed, Van, M. Anne, Linda Steg, and Peter De Jonge. 2023. 'Climate Anxiety: A Research Agenda Inspired by Emotion Research'. *Emotion Review* 15 (4): 258–262. https://doi.org/10.1177/17540739231193752.

Vaškovic, Petr. 2023. 'Philosophical Perspectives on Climate Anxiety'. In *Handbook of the Philosophy of Climate Change*, edited by Gianfranco Pellegrino and Marcello Di Paola. Handbooks in Philosophy. Springer International Publishing. https://doi.org/10.1007/978-3-031-07002-0_144.

Vaškovic, Petr, and Gabriela Vičanová. 2024. 'Anxiety, Hope and Meaning in Times of Ecological Crisis: An Existential-Phenomenological Perspective on Environmental Emotions'. *Human Studies* 47 (4): 771–791. https://doi.org/10.1007/s10746-024-09728-3.

Villalba, Bruno. 2023. 'Sobriety'. In *Handbook of the Anthropocene*, edited by Nathanaël Wallenhorst and Christoph Wulf. Springer International Publishing. https://doi.org/10.1007/978-3-031-25910-4_102.

Villalba, Bruno, and Luc Semal. 2018. *Sobriété Énergétique. Contraintes Matérielles, Équité Sociale et Perspectives Institutionnelles*. Éditions Quæ.

Vuuren, Detlef P. van, Elke Stehfest, David E. H. J. Gernaat, et al. 2018. 'Alternative Pathways to the 1.5 °C Target Reduce the Need for Negative Emission Technologies'. *Nature Climate Change* 8 (5): 391–97. https://doi.org/10.1038/s41558-018-0119-8.

WCED. 1987. *Our Common Future*. Oxford University Press.

Wiedmann, Thomas, Manfred Lenzen, Lorenz T. Keyßer, and Julia K. Steinberger. 2020. 'Scientists' Warning on Affluence'. *Nature Communications* 11 (1): 3107. https://doi.org/10.1038/s41467-020-16941-y

Williston, Byron. 2012. 'Climate Change and Radical Hope'. *Ethics and the Environment* 17 (2): 165–186. https://doi.org/10.2979/ethicsenviro.17. 2.165.

Wiseman, John. 2021. *Hope and Courage in the Climate Crisis: Wisdom and Action in the Long Emergency*. Palgrave Macmillan. https://doi.org/10. 1007/978-3-030-70743-9.

Wiseman, John. 2022. 'Hope and Courage in a Harsh Climate: From Denial and Despair to Resilience and Transformation'. In *The Palgrave Handbook of Climate Resilient Societies: Volumes 1–2*, vol. 2. https://doi.org/10.1007/ 978-3-030-42462-6_130.

Wray, Britt. 2022. *Generation Dread: Finding Purpose in an Age of Climate Crisis*. Alfred A. Knopf Canada.

Wynes, Seth, and Kimberly A. Nicholas. 2017. 'The Climate Mitigation Gap: Education and Government Recommendations Miss the Most Effective Individual Actions'. *Environmental Research Letters* 12 (7): 1–9.

CHAPTER 4

Conclusion: Learning to Live with Eco-Anxiety

Abstract This section provides an overview of the approach to eco-anxiety developed in the book. It reviews the main features of eco-anxiety, shows why it is a moral and political problem, explains what is new about this ecological emotion, and highlights possible ways to cope with it. The chapter draws on the conceptual disruption framework to explain the relative newness of eco-anxiety, and on the framework of ecological citizenship to explore a possible way to learn to live with eco-anxiety. It explains how carbon sobriety, hope, and courage are interrelated values that guide ecological citizens in their everyday lives. Finally, it suggests future areas of research on eco-anxiety from an ethical and political philosophical perspective.

Keywords Eco-anxiety · Risk · Conceptual disruption · Ecological citizenship · Hope · Courage

WHAT IS ECO-ANXIETY?

The first objective of this book was to provide a detailed conceptual analysis of the notion of eco-anxiety. Eco-anxiety is a threat-related ecological emotion oriented towards future ecological risks. It is a state of insecurity in the face of the uncertainty raised by these risks that can cause

© The Author(s) 2026
M. Bourban, *Eco-Anxiety and Ecological Citizenship*,
https://doi.org/10.1007/978-3-032-03219-5_4

paralysis and inaction but that can also lead to risk-assessment and risk-minimization behaviours. It can be an alienating and isolating feeling, especially when it is not validated, recognized, or reciprocated by those nearest and dearest, but it can also re-politicize environmental action, especially when it leads to political mobilization and coalition-building. Eco-anxiety comes in different degrees, from very low and unproblematic forms to severe and potentially pathological forms. It is not a universal emotional response to our planetary predicament, but it is felt throughout the globe and across socio-demographic categories, with children, young people, women, Indigenous people, and people directly exposed to environmental impacts being more likely to experience higher levels of eco-anxiety. It is not an anticipation of highly unlikely or exaggerated risks; it is first and foremost a lucid reaction to an accurate empirical description of global environmental changes, a fitting emotional response to a truly dangerous situation and a future that is really threatening. It is an emotional response not only to problematic uncertainty regarding the specific shape the future will take but also to problematic certainty that we are currently on a path that leads to a much more dangerous world for humans and other life forms on the planet. This tension between (problematic) certainty and uncertainty is at the core of the emotion of eco-anxiety.

Why Is Eco-Anxiety a Moral and Political Problem?

There are three main reasons why eco-anxiety represents a moral and political problem. First, it can lead to moral injury when the ecological distress felt in the face of ecological risks is further reinforced by the feeling of confusion, abandonment, and betrayal caused by the environmental indifference of adults and the inaction of policymakers. This moral injury can only be addressed if adults and policymakers recognize the validity of eco-anxious feelings and act on this recognition by implementing fair and ambitious environmental and climate policies that reflect the demands of environmental and climate justice. It is also important for policymakers to develop policies that specifically address eco-anxiety. To do so, they can, for instance: implement awareness policies based on adequate communication campaigns that present eco-anxiety as a normal emotional response to ecological threat and highlight its problematic side effects such as stress, worry, and paralysis; subsidize research projects on eco-anxiety to better understand the factors that generate high levels of eco-anxiety and find the most effective actions to regulate eco-anxiety levels; and invest in training for mental health professionals on how to deal with eco-anxiety in order to prevent its most extreme forms.

Second, eco-anxiety can also be a threat to or violation of the human right to health, understood as the right not to have one's physical and mental health threatened by the actions of individual and collective agents. This is particularly the case in situations of severe and very severe eco-anxiety, when people feel in distress every day and have minimal psychological defences to deal with their emotions and mitigate their anxiety, when they are sleep deprived and struggle to enjoy any aspect of life because of their intrusive thoughts, when they strongly believe that societal collapse will take place in the near future, and/or when they start having thoughts of suicide or even thoughts of killing their own children to save them from a violent death. This is a serious ethical problem, as the right to health is a basic right, a right that is a condition of possibility for all the other human rights. This should push environmental justice and climate justice scholars to take the impact of environmental change on mental health more seriously and to contribute more to ongoing research on eco-anxiety.

Third, eco-anxiety can also be conceived as an emotional injustice: it leads to an unfair allocation of emotional burdens. The people and communities who have contributed the least to causing environmental problems are the most vulnerable to its impacts on both physical and mental health. It is important to address this emotional injustice through a fairer allocation of emotional burdens.

Is Eco-Anxiety a New Emotion?

Eco-anxiety is not necessarily a new emotion, but there are new features in this emotional response to ecological threats. The best way to understand the newness of the notion of "eco-anxiety" is to conceive it as a result of *conceptual innovation* in reaction to the *conceptual and emotional disruption* caused by current ecological problems. Climate change, biodiversity loss, and other planetary boundary transgressions are disrupting some emotions and how we think about them, pushing us for instance to couple anxiety with eco-anxiety, grief with eco-grief, or hope with eco-hope. It is not clear if conceptual disruption is taking place in each case, but there are good reasons to think that it is the case with eco-anxiety. As we saw, anxiety and eco-anxiety share basic features, such as future orientation, uncertainty, insecurity, and epistemological responses. At the same time, they differ in at least two key aspects: the object of the emotion and its categorization as a medical condition. Let us discuss each in turn.

First, the object of eco-anxiety is ecological risks. This broad category covers three major elements: the impacts of global and intergenerational environmental problems, their effects on societies and non-human beings, and the very measures taken to address these problems and effects. The newness here is fourfold. First, the *nature* of environmental problems. Ecological risks today are no longer limited to problems of pollution and environmental degradation that are limited spatially and temporally to the size and dynamics of biotic communities and ecosystems; they are first and foremost defined by global and intergenerational disruptions in the Earth system, such as climate change and biodiversity loss. Second, the *spatial scope* of the losses and damages. The perceived ecological dangers are no longer restricted to a household, a community, or society; it can expand to civilization or humanity itself. Third, the *temporal scope* of the losses and damages. It is not only generations who are currently alive who are threatened, but future generations too. Fourth, the *subjects* of losses and damages. Eco-anxiety is not only about one's own mortality or the risks to other human beings but also the well-being of non-human beings such as sentient animals, ecosystems, and other species. This fourth element is perhaps the most original one, as it shows that eco-anxiety has both anthropocentric and non-anthropocentric components.

Second, while anxiety is a medical condition, eco-anxiety is not. As we saw, anxiety appears in the *Diagnostic and Statistical Manual of Mental Disorders* (DSM) of the American Psychiatric Association (APA), where it is conceived as an "apprehensive anticipation of future danger or misfortune accompanied by a feeling of dysphoria or somatic symptoms of tension" (APA 2000, 820). One key reason why anxiety represents a medical condition is that it can lead to mental health disorders such as phobia, social phobia, panic disorder, generalized anxiety disorder (GAD), obsessive–compulsive disorder, and post-traumatic stress disorder (PTSD) (Freeman and Freeman 2012). This does not mean that anxiety is an intrinsically negative emotion or debilitating state: it is a valuable affective state that makes us aware of potential threats and pushes us to prepare for the emergence of possible risks and dangers (Kurth 2018). However, it remains a medical condition that is a danger to one's mental health, with one-third of the adult population reporting having anxiety problems and almost one-fifth meeting the criteria for clinical disorder (Freeman and Freeman 2012, 111). This implies that in many cases, psychological therapy such as cognitive-behavioural therapy (CBT) is required, sometimes with medication such as antidepressants and anxiolytic drugs.

In contrast, eco-anxiety does not represent a medical condition in itself. It is true that eco-anxiety can be associated with mental health outcomes such as psychological distress, depression, insomnia, and stress symptoms. It can lead to impairment in day-to-day life related to eating, concentrating, work, school, sleeping, spending time in nature, playing, and/or having fun and relationships. However, most forms of eco-anxiety appear to be non-clinical. Eco-anxiety is not listed as a medical condition in the DSM, and many mental health professionals explain that it is important that eco-anxiety remains excluded from this manual. The main reason for this is that this would pathologize an emotion that arises out of a normal and healthy reaction to the severity of the ecological issues we are currently facing. To distinguish between healthy and unhealthy forms of eco-anxiety, it is essential to distinguish between different degrees of eco-anxiety. While lower and milder levels of eco-anxiety are not a direct threat to mental health, beyond a certain threshold, levels of eco-anxiety become too severe and can lead to pathological states. But, once again, that does not mean that eco-anxiety should be conceived as a medical condition.

These two key differences between anxiety and eco-anxiety show that the latter represents a relatively new emotion. The newness of eco-anxiety is perhaps best understood from the perspective of conceptual disruption and conceptual innovation. The new kind of ecological problems we are facing today create a conceptual disruption by challenging our theoretical bearings and thereby pushing us to rethink certain concepts we use to think about mental health or well-being. More specifically, contemporary ecological risks create a conceptual gap where our existing conceptual repertoire fails to identify the key features of an already existing concept (in this case, anxiety).[1] This gap is wide enough to create the need for a

---

[1] I draw here on the account of conceptual disruption provided by Ibo van de Poel et al. (2023) and Jeroen Hopster and Guido Löhr (2023), but I adapt it, as they restrict their account to disruption caused by new and emerging technologies. The role of technologies is also important here, as they play a key role in the creation and development of current environmental problems and pose new risks when used to address them, but the primary cause of conceptual disruption discussed here is environmental problems themselves. Here, I follow Allen Thompson (2010, 44), who explains that climate change is altering our

more suitable concept: this is why reinterpreting the notion of "anxiety" is not sufficient; rather we need to engage in conceptual innovation and use the notion of "eco-anxiety" instead. Again, there is continuity between the two concepts, and this is why it is important to keep "anxiety" in the new concept, but at the same time, it would be misleading to reduce eco-anxiety and climate anxiety to anxiety tout court.

WHAT CAN BE DONE TO DEAL WITH ECO-ANXIETY?

Finding ways to cope with eco-anxiety was the second objective of this book. We saw that many coping strategies are possible. Although the psychological and psychiatric approach is relevant, especially in cases of severe and very severe forms of eco-anxiety, this approach needs to be complemented because of the problem of limited access to mental health services and resources and the risk of medicalizing eco-anxiety. This book proposes ecological citizenship as a promising normative approach to cope with eco-anxiety. Ecological citizenship makes it possible to harness the positive sides of eco-anxiety by acting as a medium between apprehensiveness of environmental threats and environmental action. Eco-anxiety leads to epistemic behaviours, such as information gathering, which is concerned with good or accurate decision making, in the context of lifestyle changes, for instance. Ecological citizenship can reinforce such decision-making processes and changes in lifestyle as it is related to citizenship activity both in the public sphere where environmental policies are designed and implemented and in the private sphere of individual choices in terms of consumption and family size. Ecological citizenship questions the dichotomy between the private and the public spheres, as private acts, such as household energy use, choice of diet, or the size of one's family now have public consequences in that they contribute to environmental impacts at the collective—and even global—level. This is why being an ecological citizen does not just mean being a certain kind

understanding of normative notions such as virtues: "impending radical changes in global climate will likely precipitate significant changes in the dominant world culture and then consider how these changes could alter the moral landscape, particularly culturally thick conceptions of the environmental virtues. Our understanding of environmental virtues may be moored to cultural and environmental pilings that are unstable and we should begin to consider the kinds of character traits best suited to radical change".

of voter, elector, or activist, but also being a certain kind of consumer, producer, or worker.

Ecological citizenship is also useful as a way of addressing the problematic sides of eco-anxiety. Being eco-anxious can be a very isolating experience. This is especially the case for people suffering from moral injury, such as children having to face unresponsive and indifferent adults and government officials. Ecological citizens belong to a moral and political community based on the commitment to make environmental protection the centre of lifestyle and policy choices. By joining this community of like-minded people, eco-anxious people who feel isolated, alienated, or ostracized move from a path that can ultimately lead to despair, depression, and paralysis to a constructive path of hopeful and courageous actions reflecting carbon sobriety.

What are the core values of ecological citizens? The first is carbon sobriety, which is an alternative to consumerism, a form or moderation, a way of living well with less. Conceived as a green virtue, it represents an acquired and stable disposition to enjoy consuming less, to lead a flourishing life by living more soberly when it comes to emitting activities, especially high-emitting activities. It is a mindset, a state of character that informs everyday actions, a fundamental lifestyle choice. One of the best ways to express carbon sobriety is to implement, whenever reasonably possible, actions that have a high impact in reducing one's environmental footprint.

The second value is hope, but not any form of hope: to avoid the pitfalls of wishful thinking, disappointment, and distraction, priority should be given to cultivating good hope and radical hope. Good hope draws on the relational feature of eco-anxiety; it is a form of social hope that encourages hoping together with responsive people. It transforms eco-anxiety into motivation to mitigate ecological risks, rather than letting these risks become a source of isolation, despair, and paralysis. Radical hope is an acceptance of the possibility of ecological collapse, an openness to the dangers that come with a radically uncertain future in which familiar cultural frameworks are at risk of dissolving, and a belief in the possibility of meaning beyond the loss of the world as we know it. Good hope and radical hope both make it easier to live with eco-anxiety without being overwhelmed by it.

The third value is courage. Living beyond planetary boundaries and in the range of tipping points requires not only hope but also courage in the face of the possibility that the world may well become more hostile for humans and other species, which will threaten both human and non-human flourishing. Courage is not fearlessness but the capacity to defend a worthy cause despite the risks incurred in doing so. Courage helps to translate carbon sobriety and hope into action by providing the willingness to protect environmental value in spite of the physical and social risks arising from environmental actions at the individual and collective levels. Courage is an environmental virtue, especially when it becomes climate courage, that is, an acquired and stable disposition to act in spite of the fact that we live in a state of planetary emergency.

Future Research

This book provides one of the first philosophical perspectives on eco-anxiety from an ethical and political philosophy angle. Many topics discussed here need to be further investigated, including the links between the mood state and the emotion state of eco-anxiety, the connection between eco-anxiety and other components of the anxiety culture, such as techno-anxiety, the relation between eco-anxiety and other ecological emotions, such as eco-grief and eco-despair, the differences and relations between the epistemic components of eco-anxiety, and the policies needed to bring eco-anxiety to the forefront of the political agenda as well as the measures to provide mental health professionals with the required means to identify and address different forms of eco-anxiety. The conceptual analysis in Chapter 2 was meant to clarify the contours of the notion of eco-anxiety, but this work needs to be continued by drawing on the rapidly increasing empirical research on the topic, typically in terms of cognitive and behavioural manifestations of this ecological emotion—the definition I provided is a working definition and therefore needs to be further developed. The framework of ecological citizenship discussed in Chapter 3 represents an appropriate response to eco-anxiety as it supports the radical shifts in lifestyles and policy that are required to properly address current environmental problems, but other coping strategies are possible and worth pursuing, starting with Indigenous and local perspectives on eco-anxiety and other related ecological emotions.

As global environmental problems will continue to worsen and their impacts will become more severe, frequent, and widespread, levels of eco-anxiety will probably continue to increase throughout the globe and across socio-demographic groups. Threat-related emotional responses to ecological risks will become increasingly fitting as new people and communities must face the consequences of planetary boundaries being transgressed and the first tipping points being crossed. In these circumstances, the most worrying response is not that of eco-anxiety; it is that of eco-despair, which is based on the belief that the object of eco-hope has become impossible to reach or is, at least, extremely unlikely to be reachable. Eco-anxiety and eco-hope go hand in hand; they are both future-oriented reactions to uncertainty. Once the element of doubt is replaced by the conviction that alternatives to dangerous scenarios such as the Hothouse Earth pathway are no longer possible or extremely unlikely, resignation and paralysis replace hope and courage. One of the most important responsibilities of our time, which falls especially on those who hold political, economic, and social power, is to ensure that eco-anxiety does not fade into eco-despair.

References

APA. 2000. *Diagnostic and Statistical Manual of Mental Disorders: DSM-IV-TR.* American Psychiatric Press.

Freeman, Daniel, and Jason Freeman. 2012. *Anxiety: A Very Short Introduction.* Oxford: Oxford University Press.

Hopster, Jeroen, and Guido Löhr. 2023. 'Conceptual Engineering and Philosophy of Technology: Amelioration or Adaptation?' *Philosophy & Technology* 36 (4): 70. https://doi.org/10.1007/s13347-023-00670-3.

Kurth, Charlie. 2018. *The Anxious Mind: An Investigation into the Varieties and Virtues of Anxiety.* Cambridge: MIT Press.

Thompson, Allen. 2010. 'Radical Hope for Living Well in a Warmer World'. *Journal of Agricultural and Environmental Ethics* 23 (1–2): 43–59. https://doi.org/10.1007/s10806-009-9185-2.

Van De Poel, Ibo, Jeroen Hopster, Guido Löhr, Elena Ziliotti, Stefan Buijsman, and Philip Brey. 2023. '1: Introduction'. In *Ethics of Socially Disruptive Technologies*, 1st edn, edited by Ibo Van De Poel, Lily Eva Frank, Julia Hermann, et al. Open Book Publishers. https://doi.org/10.11647/obp.0366.01.

INDEX